Manfred Powis Bale

Stone-Working Machinery and the Rapid and Economical Conversion of Stone

Second Edition

Manfred Powis Bale

Stone-Working Machinery and the Rapid and Economical Conversion of Stone
Second Edition

ISBN/EAN: 9783337864996

Printed in Europe, USA, Canada, Australia, Japan

Cover: Foto ©Suzi / pixelio.de

More available books at **www.hansebooks.com**

STONE-WORKING MACHINERY.

WORKS BY THE SAME AUTHOR.

WOOD-WORKING MACHINERY; its Rise, Progress, and Construction. With Hints on the Management of Saw Mills and the Economical Conversion of Timber. Illustrated with Examples of Recent Designs by leading English, French, and American Engineers. Second Edition, Revised. Crown 8vo, 9s. cloth.

"The most comprehensive compendium of wood-working machinery we have seen. The author is a thorough master of his subject."—*Building News.*

SAW-MILLS; THEIR ARRANGEMENT AND MANAGEMENT, AND THE ECONOMICAL CONVERSION OF TIMBER. (Being a Companion Volume to "Wood-Working Machinery.") With numerous Illustrations. Second Edition, Revised. Crown 8vo, 10s. 6d. cloth.

"Will be found of much value by that special class of readers for whose information it is designed. We have pleasure in recommending the book to those about to construct or to manage saw-mills."—*Athenæum.*

STEAM AND MACHINERY MANAGEMENT. A Guide to the Arrangement and Economical Management of Machinery, with Hints on Construction and Selection. Crown 8vo, 2s. 6d. cloth.

"Gives the results of wide experience."—*Lloyd's Newspaper.*

PUMPS AND PUMPING; A Handbook for Pump Users. Being Notes on Selection, Construction, and Management. Third Edition, Revised. Crown 8vo, 2s. 6d. cloth.

"Thoroughly practical and simply and clearly written."—*Glasgow Herald.*

CROSBY LOCKWOOD AND CO.,
7, STATIONERS'-HALL COURT, E.C.

STONE-WORKING MACHINERY

AND THE

RAPID AND ECONOMICAL CONVERSION OF STONE.

WITH

HINTS ON THE ARRANGEMENT AND MANAGEMENT
OF STONE-WORKS.

BY

M. POWIS BALE,

M. INST. M.E., A.-M. INST. C.E.,

AUTHOR OF "WOODWORKING MACHINERY, ITS RISE, PROGRESS, AND CONSTRUCTION;"
"SAW MILLS, THEIR ARRANGEMENT AND MANAGEMENT;"
"A HANDBOOK FOR STEAM USERS," &C.

With Numerous Illustrations.

Second Edition, Revised and Enlarged.

LONDON:
CROSBY LOCKWOOD AND SON,
7, STATIONERS' HALL COURT, LUDGATE HILL.
1898.

[All rights reserved.]

BRADBURY, AGNEW, & CO. LD., PRINTERS,
LONDON AND TONBRIDGE.

PREFACE.

The following pages have the honour, and at the same time, the disadvantage of being the first work published—in this country—on the Conversion of Stone by Machinery. The disadvantage to an Author in not having books of reference on the subject on which he treats is sufficiently obvious; and some of the shortcomings of the present work will, the Author trusts, be kindly attributed to this cause. Although the conversion of stone is one of the most ancient of all the mechanical arts, its conversion by means of machinery—in an advanced form—is quite of modern origin. Much has been done during recent years to develop this branch of engineering, and although many obstacles have presented themselves, very marked success has already been attained. Ample scope, however, remains for still further improvements.

In erecting stone-working machinery it is of the highest importance that its principle of working be adapted to the nature of the stone to be operated on; and by judicious employment it can, without doubt, be made to effect an immense saving over hand-labour.

The Author having had considerable experience in the construction of stone-working machinery, trusts the hints he gives may be found of service to those about to erect such machinery, as well as to others who have it already in operation.

APPOLD STREET, LONDON, E.C.
September, 1897.

CONTENTS.

CHAPTER	PAGE
I.—Introductory	1
II.—Stone suitable for Machine Conversion	8
III.—Arrangement of Stoneworks for General Purposes	16
IV.—Stone-sawing Frames	22
V.—Stone-sawing, &c., with Diamond Points	36
VI.—Circular Saws for Cutting Stone	40
VII.—Stone-dressing and Planing Machines	52
VIII.—Stone-moulding Machines	77
IX.—Rubbing or Surfacing Beds	89
X.—Stone Recessing and Moulding Machines	98
XI.—Sculpturing Machinery	103
XII.—Polishing Beds	109
XIII.—Turning Lathes	115
XIV.—Stone-breaking Machinery	118
XV.—Slate-working Machinery	124
XVI.—Miscellaneous Machinery for Working Stones	135

CHAPTER	PAGE
XVII.—CUTTING TOOLS	138
XVIII.—NOTES ON MANAGEMENT OF A STONE WORKS	149
XIX.—STONEWORKING: HAND LABOUR VERSUS MACHINERY	155
XX.—NOTES ON MASONRY, RECIPES, &C.	162
APPENDIX A.	170
APPENDIX B. TECHNICAL TERMS	173

INDEX . 178

LIST OF ILLUSTRATIONS.

FIG.		PAGE
1. Plan of Stoneworks for General Purposes		17
2, 3. Horizontal Stone Sawing Frame .	. *To face*	35
4, 5, 6, 7, 9, Cutting Tools .	. 44, 45,	48
8. Stone Ridges		47
10. Cutter-stocks	.	58
11. Horizontal Barrel : Stone-dressing Machine	*To face*	62
12, 13, 14, 15, Chucks and Cutting Tools .	. 67, 68,	69
16. Machine for Top-dressing Hard Stones .	*To face*	70
17. Machine for Side-dressing Stones	. .	72
18. Horizontal Planing and Moulding Machine .	*To face*	73
19. Tool and Tool Holder . .	.	74
20. Trier's Patent Vertical Stone-dressing Machine	*To face*	76
21. Section of Steel Tool		80
22. Duplex Stone-moulding Machine	. .	81
23, 24, 25, 26, Tools and Tool Holders	82,	83

b

LIST OF ILLUSTRATIONS.

FIG.		PAGE
27. Vertical Barrel Stone-moulding Machine (*Back Elevation*)	⎫	
28. Vertical Barrel Stone-moulding Machine (*End Elevation*)	⎬ *To face*	85
29. Vertical Barrel Stone-moulding Machine (*Plan*)	⎭	
30. Horizontal Stone-moulding Machine .	*To face*	88
31. Horizontal Rubbing Beds .		96
32. Hand-power Rubbing Disc		97
33. Tool for Cutting Grooves		100
34. Horizontal Polishing Machine	.	110
35. Vertical Polishing Disc .	.	112
36. Hand Polishing Rubber	.	113
37. Stone-breaking Machine . .		121
38. Cylinder Drill	.	137
39. Turning Tool Holder .		147
40. Section of Moulding . .		159
41. Circular Stone Sawing Machine. .		171

STONE-WORKING MACHINERY.

CHAPTER I.

INTRODUCTORY.

One of the most ancient of all the mechanical arts is that of working and preparing stone for building purposes. This is abundantly proved by the remains discovered in Egypt, Greece, and other countries; and that the art of the stonemason was also one of considerable importance many centuries ago, can hardly be doubted, from the constant reference made to it in Holy Writ and other ancient histories. Notwithstanding the important part the stonemason has played for centuries in the construction of the great monuments and buildings of the world, it is a somewhat astonishing fact that, till within comparatively recent years, no improvement of importance —so far as at present can be gathered—has been effected in the processes of stoneworking, and the ancient mode of sawing stone by means of a plate of iron stretched in a frame, and reciprocated horizontally by the hands of a sawyer seated before it, is still largely practised.

According to Manetho, Sæsorthus introduced the art of building with hewn stone; but the Phœnicians are probably entitled to the credit of being the first users of the stone-saw, for we read that they erected the Temple of King Solomon of stone sawn within and without. The first buildings, however, of stone to which a date has been assigned, are the Pyramids of Ghizeh.

According to Diodorus, the great wall surrounding the city of Nineveh was 100ft. high, and so broad that three chariots might drive abreast. Xenophon says that this wall, up to 50ft. high, was composed of blocks of fossiliferous limestone, smoothed and polished on the outside. A tower is also in existence composed of large blocks of marble, cut with great exactness, and joined together without mortar or cement of any kind; the roof consists of four large slabs of stone, reaching entirely across from side to side, and measuring about 24ft. in length, and 6ft. in width, and from 18in. to 3ft. in thickness. The slabs are cut to slope each way from the diagonal lines, and were originally carefully clamped together with iron clamps. A large portion of the masonry of the palace at Khorsabad was of stone, square hewn and of great size, laid dry and backed with bricks.

The obelisks, monoliths, and colossi of Ancient Egypt now in existence prove, without shadow of doubt, that the art of quarrying stone was well understood by the Egyptians thousands of years ago. The obelisks transported from the quarries of Syene, at the First Cataracts, to Thebes and Heliopolis, vary from 70ft. to 93ft. in length. The largest monolithic obelisk in Egypt is that at Karnak, and is calculated to weigh 297 tons. The statue of Rameses II., when entire, is stated to have weighed 887 tons, and no monuments of bygone ages have excited more wonder, controversy, and speculation, as to the manner of their quarrying, transport, and erection, than those of Ancient Egypt.

Ancient paintings taken from Thebes illustrate the

various operations of bevelling, squaring, and chiselling stone. The straight-edge used, appears to be a taut cord, and the chisels and mallets much like our own. In a tomb at Thebes workmen are represented mounted on a scaffold, and working at a sitting Colossus of granite, and in the large platform of the Temple of Baalbec are to be found worked stones 64ft., 63ft. 8in., and 63ft. long, with a height and width of 13ft.

Of the early history of stone-working by what may properly be called machinery, very little is known. Ausonius speaks of water-mills for cutting stone being erected as early as the fourth century, on the small river Roeur in Germany; but what these mills consisted of is not stated. It is recorded, however, that the earliest forms of stone saws consisted of flakes of flint imbedded in a wooden blade, and held in position by a composition of bitumen.

Pliny conjectures that the saw-mill for cutting stone was invented in Caria; at least he knew of no building covered with marble of greater antiquity than the Palace of King Mausolus, at Halicarnassus. This was erected, according to one authority, 350 B.C. This edifice is celebrated by Vitruvius for the beauty of its marble: and Pliny gives an account of the kind of sand used for cutting it; for "it is the sand," he says "and not the saw, which produces that effect." The latter presses down the former, and rubs it against the marble; and the coarser the sand is the longer will be the time required to polish the marble which has been cut by it. It appears that stones of the soapstone-rock kind, which are much softer than marble, were sawn at this period; but it is recorded that far harder glassy kinds of stone were sawn or cut also; for we are told of the discovery of a building which was encrusted with cut agate, carnelian, lapis-lazuli, and amethysts.* The ancient Egyptians doubtless practised to a considerable extent the art of stone-cutting, as many arched vaults of cut stone were found amongst the ruins of Nineveh, and some are still remaining at Thebes. The magnificent masonry of the Parthenon at Athens, and many other ancient Greek temples, shows that stone conversion was in a considerable state of development in several countries many centuries ago, and it is generally allowed that in the remains of ancient Greece the best models of unmixed Doric, Ionic, and Corinthian architecture are to be found.

Crassus is credited with being the first Roman who embellished his house with sawn marble about 90 B.C., and it is recorded that several palaces of the Cæsars were made of it. Cornelius Nepos states that Mamurra, at a little later date, was the first to use marble in this way, whilst Artemis of Caria states that sawn marble was used for building purposes several hundred years before these dates.

In Italy, the dome of the Pantheon at Rome, which is a hemisphere of 139 ft. dia., and the vaulted roofs of the halls of the Baths of Diocletian and Caracalla, may be cited as simple but grandly-executed specimens of Early Italian stonework. The dome of St. Peter's at Rome, built from the designs of Michael Angelo at the close of the 16th century, affords a great example of the more modern Italian school of stone-working. The ancient city of Syracuse, in Sicily, exhibits remains which conclusively prove, both from its buildings, and the enormous disused stone quarries which exist in the neighbourhood, that stone quarrying and stone-working were practised here in ancient days to a very large extent.

* See Jannon de S. Laurent's treatise on the cut stones of the ancients, in Saggi di Dissertazioni nella Acad. Etrusca di Cortona, tom. 6, p. 56.

One of the finest pieces of polished marble-work to be seen, perhaps, in the world, is the Taj, about three miles from Agra. It consists of an octagon tomb, surmounted by an egg-shaped dome 70 ft. in circumference, and four minarets 150 ft. in height. These are all built of pure white polished marble, inlaid with one large flower mosaic, composed of different coloured stones. The screen of the tombs is divided into compartments and panels, and runs round marble cenotaphs that lie within. This screen is of the purest marble, so pierced and carved as to look like a high fence of exquisite lacework, but is represented by those who have seen it as being far more refined and beautiful. Along the panels are wreaths of flowers composed of lapis-lazuli, jasper, heliotrope, chalcedony, carnelian, &c.

As far as this country is concerned, stone-working was, of course, practised to a certain extent during the Saxon, Norman, and following periods, and in the time of Henry I. the choir of Canterbury Cathedral was paved with marble, but it cannot be held that stone-working arrived, in England, at any advanced state of development much before the 16th century.

As regards the literature of stone-working, with the exception of the old works by De Lorme (1568) and Halfpenny (1725), very little record has been kept of the early history of the subject. A few modern works have, however, been published; but no treatise, as far as we are aware, dealing with stone-conversion by machinery: it is the aim, therefore, of the author, in a measure, to supply this deficiency.

It may here be asked, What can machinery do in the way of stone-conversion at the present time? This may be briefly summarised as follows :—Stone may be sawn, dressed, squared, faced, and polished, architrave mouldings, cornices, ovolos, pilasters, astragals, ogees, scotias, strings and other straight, undercut, and curved mouldings may be shaped and finished in every way superior to, and at an immense saving over, hand-labour. All the heavy work in small mouldings, panels, recesses, &c., can also be worked. Hard flagstones, and even granite, may be dressed; landings, copings, steps, channellings, &c., may be tooled, granite turned and polished, and many other operations may be performed too numerous to recapitulate here. Although much has been done during recent years, as regards the introduction of machinery for working stone, ample scope still remains for inventors in this direction.

The adaptation of machinery to common uses in these degenerate days of strikes, high wages, and short hours, is, without doubt, increasingly necessary to promote the commercial prosperity and progress of a nation. Mr. Ruskin's Utopian ideas may, as sentimental theories, have something to commend them, but cannot for a moment be practically entertained, and engineering science must of necessity ever play a more and more prominent part in the economy of production. It has been many times maintained that the introduction of machinery bears heavily on the working classes by dispensing with manual labour: this fallacious argument has, however, been sufficiently disproved by the fact that the introduction of labour-saving machinery has not lessened, but rather raised, the wages of skilled artisans, as it is found the cheaper production creates the greater demand. It has also been argued that machinery has damaged art progress by reducing the production of skilled handicraft to a mere dead mechanical level. This, however, with men who have any art instincts in their composition, is not the case; nor should it be with any, but should rather act as an incentive for the workman to attain to a higher art knowledge, both theoretical and practical, the results of which no mechanical contrivance can rob him of. At the same time, in stone-working and decorative construction, we take it that machinery should be a powerful hand-maiden to art workmanship, by doing much of the heavy, laborious preparation of the crude material, leaving the skilled workman to give the finishing touches to the whole. We do not claim for stone-working machinery that it can at present produce the "storied windows richly dight," and possibly it is as well in an art sense that this is so; but we do claim that it may, by being judiciously employed, be made not only most remunerative, but at the same time a help and assistance to art, and not a hindrance, as is asserted by what may be called the ultra-sentimental school.

CHAPTER II.

STONE SUITABLE FOR MACHINE CONVERSION.

The stones used in building may be classed under four general divisions—viz., Limestones, Sandstones, Granites, and Slates. Limestone, including magnesian limestone, and the oolites in their different varieties, is undoubtedly the most important class of stone used in building operations, as it combines considerable durability with tolerable facility in working. Limestones are composed of carbonate of lime, and the carbonates of lime and magnesia mixed with foreign matters in variable proportions, and often with oviform bodies called oolites.

Sandstones are principally silicious, and are generally laminated, especially if they contain mica; they are generally composed of either quartz or silicious grains, cemented by silicious, argillaceous, calcareous, or other matter. Granite is of granular structure, and owing to its crystallisation and hardness, is extremely difficult to work; it contains, in varying proportions, quartz, felspar, and mica, and the variety of the proportions in which these are found gives the various colours.

Stones used in building construction are generally known either by the names of the places from which they are quarried or from the chief or some special ingredients of their composition. The term "freestone" is applied indefinitely to that kind of stone which can be wrought with a mallet and chisel, or cut with a saw, and it includes the two great divisions of limestones and sandstones.

Stones used for conversion into mouldings, &c., by machinery, should be free and even in texture and hardness; very tender stones are unsuitable, as also those stones which contain much shelly fossil deposit, especially if the shells should be hard and crystalline; but from a casual examination of a stone, owing to the varied nature of its constituents, it is almost impossible to tell whether it will work satisfactorily; the best plan is to try it on a machine. Stones that are much laminated in their structure are not suitable for deep-cut mouldings; they may be used for facings, but should be dressed and placed so that their planes of lamination may be in a horizontal position, or in the same position they held in the quarry. Should they be placed in a vertical position, decomposition will take place in flakes.

The stones of the nature of Bath, Portland, Caen, and York, and most kinds of freestones, can be readily worked by machinery. For outside moulding, the stone selected should be clean and bright, and uniform in colour, which generally implies uniformity of structure. Stones that absorb much moisture, or are much affected by frost, or the carbonic acid of the atmosphere, should be avoided. To ascertain whether a stone is especially susceptible to imbibe moisture, immerse it in water for a few days, and weigh it carefully before and after immersion. The hardest stones are not always the worst to work, those of a rough gritty nature being in many cases more difficult to operate on.

If the stones used are somewhat tender, to avoid

breaking the arrises when they are being dressed or moulded, revolving cutters will be found better for this kind of stone than fixed ones, which take a scraping cut, and are more likely to "pluck" the stone—that is, force pieces out of it below the surface. Hard limestones, or even granite, may be dressed to a surface, but are extremely difficult to mould. The softer kinds of stones, such as Bramley Fall, Newcastle Grit, &c., can, with a little care, be readily planed or moulded. Stone that is crumbly or short in the grain is unsuitable. Hard grit stone or flag paving, magnesian limestones, and oolites, may be dressed to a plane surface.

In selecting stones, it should be borne in mind that the hardest stones do not necessarily possess the greatest toughness or tenacity; those which contain a large amount of silex are usually very brittle; and, therefore, difficult to work; they are, however, extremely durable. Argillaceous stones, which usually contain a considerable amount of iron and argil or clay in their composition, are, as a rule, unsuitable for machine conversion, as, after exposure, the surface generally shivers off. This is attributed to the action of the oxygen of the air, in combination with the iron contained in the stone, causing the latter to swell and shiver away on the surface.

Alabaster may be readily worked into mouldings if carefully handled. Two varieties of stone are called alabaster—viz., gypsum and stalactitical limestone, which is similar to granular limestone. The gypsum alabaster is a sulphate of lime; but, being very soluble in water, is entirely unsuited to outdoor work. Slate, unless it be rotten or shaly in its character, can be worked by machinery with facility.

After the stone is obtained from the quarries, and before use, it should be exposed to the sun and winds, so that the damp or "quarry sap" is dried out of it, and the stone thoroughly seasoned, before being placed in a building; the length of time necessary for this seasoning process, of course, would vary according to the nature of the stone and situation of the quarry; but from six to twelve months is usually sufficient. A considerable quantity of freshly-quarried stone is often, without doubt, worked up and put into buildings, and with some kinds of stone possibly without much detriment: but with the softer stones, such as Bath, this practice is very detrimental to its wearing powers. In some quarries during the winter the stone is dried by means of coke-fires burning in open grates; this drying is, however, only partial, and not by any means so effectual as the natural air, or summer-drying process.

No certain rules can be laid down for testing the nature of the various stones; this resolving itself chiefly into a matter of practical experience. The following tests are, however, sometimes practised. A piece of stone is taken fresh from the quarry and squared up, to give sharp edges for examination of its texture by a lens, and to test its hardness by means of a knife or chisel. Those stones that are easily cut or scratched effervesce rapidly by dilute hydrochloric acid, and the durability of stone may, in a degree, be ascertained by the rapidity or slowness of effervescence. Most limestones are easily scratched, and effervesce rapidly under acid; magnesian limestones and dolomite do not effervesce quickly; and some, like gypsum sandstone, do not effervesce at all. The texture of a stone can be tested by chipping an angle. The blowpipe is occasionally used for testing stones, and those which contain much carbonate of lime are soon reduced to quicklime; some yield copper, lead, &c., and are more or less fusible under the action of strong heat. By saturating stones with water, and exposing them to the action of cold as produced by freezing mixtures, valuable results as to the resistance of various stones to action of the weather, may be obtained. Such a plan, however, is difficult and tedious, and a Mr. Brard, after many trials, found that sulphate of soda very closely resembled in its action on stone, the freezing of water. A saturated solution of sulphate of soda was made in cold water, the stone was immersed, and the solution boiled for half an hour; the stone was then taken out, and put into a plate with a little of the solution. It was then left for 24 hours in a cool place, and was covered with a snowy efflorescence, the liquid having disappeared, either by evaporation, or by absorption. The stone was then sprinkled gently with cold water, until all the saline particles had disappeared from the surface. After this first washing, the surfaces of the stone were covered with detached grains, scales, and angular fragments, and the stone being one that was easily attacked by the frost, the splitting of the surfaces was very decided.

One writer on the subject says:—The causes of durability of stone, and the correspondent causes of failure and decay, are either chemical or mechanical, and may be described either as decomposition or disintegration. Durability also depends much on the power of resistance to wear.

Decomposition is caused by some of the elements of the stone entering into such new combinations with water, gases, or acids as render them soluble either by the air or water. Granite, though the hardest of building stones, is liable to serious decomposition when the feldspars are alkaline, and will unite with water or acids. Some qualities of this stone are rapidly decomposed by the sea, and the same is the case with many of the limestones

Stones containing iron are also liable to decay. In its native state it is usually in a low state of oxidation, and is liable to be acted upon by additional quantities of oxygen or carbonic acid. This sort of decomposition is much increased by being alternately wet or dry, or by frequent changes of temperature. Stones, however, containing iron in a high state of oxidation as rosso antico, porphyry, &c., do not readily become decomposed. The most curious discovery of modern times is with regard to the magnesian limestones and dolomites. These were chosen for the Houses of Parliament on account of their durability. The work at Southwell Minster, 800 years old, bears every mark of the tools to the present day, and every circumstance seemed to justify its selection. It appears, however, that magnesia has a great affinity for sulphur; and the consequence is, the sulphurous acid which is present in such quantities in the smoke of London, has already caused serious decomposition in that building, as well as in the Lincoln's Inn Hall. This acid has also so much effect on the softer limestones, that the fronts of several important buildings, Buckingham Palace among the rest, have been obliged to be painted to save them from decay.

Disintegration, as has before been said, is the separation of parts of stone by mechanical action. The chief cause is the freezing of minute portions of water which get into pores, or fissures, or between the laminæ of stones, and swell slowly as crystals of ice are gradually formed, and consequently burst open the pores, or split the grain of the stone. The south sides of buildings, in northern climates, suffer more than others, as their surface becomes thawed and filled with wet in the day, and frozen again at night, more frequently than the others. A very common error in the present day is the not taking care to set stones with their laminæ or grain, or as the workmen call it "bed," in a horizontal direction. If work be "face-bedded," the action of the weather will cause the laminæ to scale off one after the other, just as the leaves of a book fall over, if the volume be placed on its back in an upright position.

Resistance to wear is another obvious cause of durability, but this depends rather on the toughness, than the mere hardness of material, a quality often attended with brittleness, as also on its situation. The crushing weight of Portland is about 10,000, while that of York is about 12,000, or one-fifth more, but in many situations Portland steps will last much longer than York. Again, the crushing weight of Peterhead Granite is about 18,000, or not quite double that of Portland; whereas if used as street paving, it would out-last six sets of the latter.

The most important building-stones found in Great Britain are the following:—

Granites, produced chiefly in Cornwall, Devonshire, Leicestershire, Aberdeenshire, and in Wicklow and Carlow, the Channel Islands, Isles of Lundy, Anglesea, and Man.

Porphyries, Syenites, Elvans, which are obtained from Cornwall, Devonshire, Leicestershire, and many parts of Scotland and Ireland.

Sandstones, the chief quarries of which are in Yorkshire, Derbyshire, Shropshire, Herefordshire, Monmouthshire, Surrey, &c., and in several of the Scottish counties. The Derby Dales, Cragleith, and other celebrated stones belong to this class.

Millstone Grit is found largely in Derbyshire, Yorkshire, and in most of the coal-producing districts.

Magnesian Limestones or Dolomites are chiefly obtained

from Yorkshire, Durham, Northumberland, Derbyshire, and Nottinghamshire.

Oolites, found at Bath, Portland, Ancaster, Ketton, and other places.

Limestones.—These are very varied; the Purbeck marble, the Derbyshire marbles, the Lias beds, the Devonian limestones, and the mountain limestones being examples.

Slates.—These are obtained in North Wales, Devonshire, and Cornwall, and in some parts of Scotland and Ireland.

Kentish Rag is obtained chiefly near Hythe and Folkestone.

CHAPTER III.

ARRANGEMENT OF STONEWORKS FOR GENERAL PURPOSES.

In arranging works for the conversion of rough stone into the various finished forms used in building construction, the selection of a suitable site, and the general arrangement of the building and machinery is a matter of very great importance in securing economy of working. No arbitrary rules can, of course, be laid down; but advantage should always be taken of the site with reference to rail, road, or water carriage, as much money may be lost through unnecessary haulage. Each site must be judged on its merits. If the stone converter is a quarry owner, and the quarry is tolerably easy of access, it may be well to erect the stoneworks contiguous to the quarry, so that the rough stone may immediately be reduced in size and rendered more portable. If, however, a tramway is laid to convey the stone to a railway, a mill near to or alongside the railway line has some advantages. Again, should water-power be procurable in the neighbourhood, by all means take advantage of it, if anyway possible, as it will pay much better to run a tramway to the water, than to use steam or other power. Unless the works are very extensive, in which case it may pay to lay down a railway of the ordinary 4ft. 8½in. gauge, a tramway of 2ft. gauge will be found suitable: this can be laid with cross sleepers and 16lb. bridge rails, at a cost of about £475 to

ARRANGEMENT FOR GENERAL PURPOSES.

£500 per mile, should no difficulties of site occur. A 2ft. tramway will also be found more convenient than a wider one for traversing the quarry, machine shop, and finishing sheds. The size and shape of the building, containing the various stone-working machines, should be varied according to the nature of the work to be carried on; but, speaking generally, it should be as open as possible, so that a free current of air may carry away any dust floating in the atmosphere. Our plan (Fig. 1) represents stoneworks suitable for general purposes; length 110ft. by 50ft. wide. This consists briefly of a pair of lean-to sheds facing each other, with an overhead traveller for lifting the blocks of stone and placing them

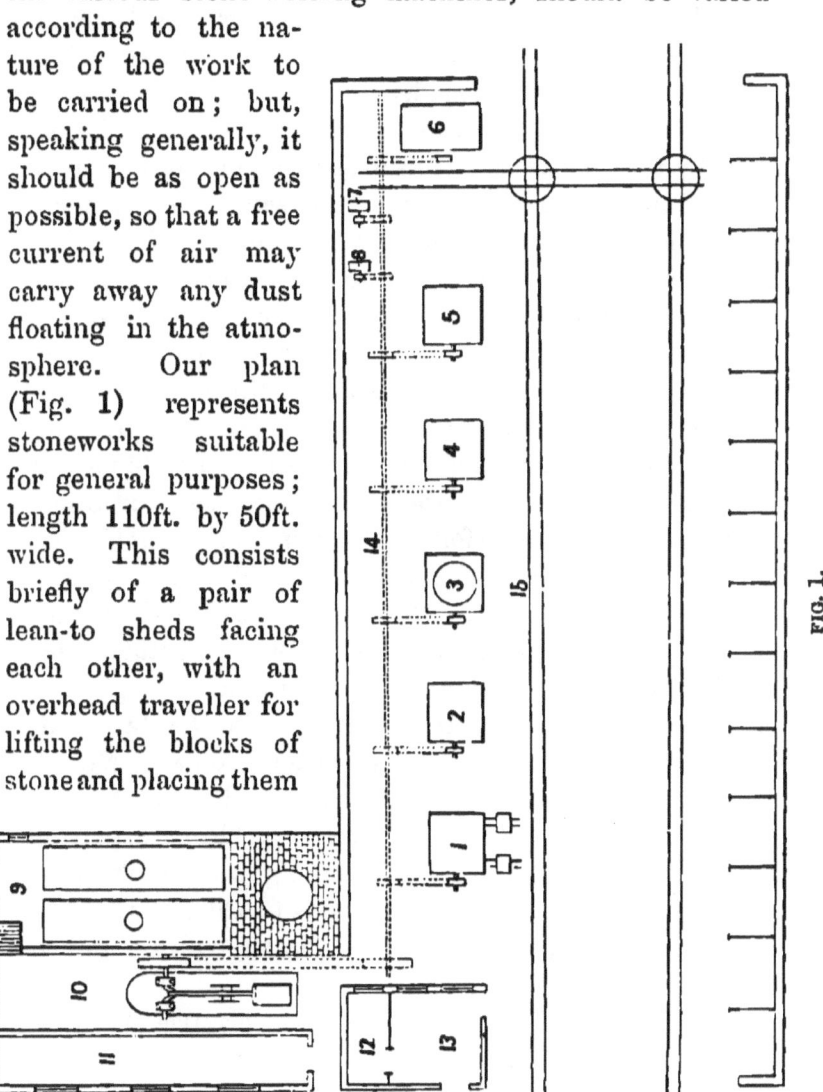

FIG. 1.

on, or removing them from the various machines passing between them; if first cost is no object, however, the whole area is best covered in. In this case the buildings should be lofty, and the masons' berths arranged on the opposite side of the mill,* and at a sufficient distance away that the chips and dust from their chisels could not penetrate the bearings, &c., of the stone-working machines, or they will rapidly deteriorate. The various bearings should in all cases be protected from the dust made by the machines themselves; but on these points we shall have something further to say. In our plan we suppose steam to be the motive power, and this we arrange in a building (Nos. 9 and 10) which is entirely separated from the works. The engine represented is a 25h.p. horizontal high pressure or compound, with a pair of 15h.p. cross-tubed Cornish boilers. No. 1 represents a horizontal double stone sawing frame. No. 2 represents a circular sawing machine, for converting large blocks of stone as they come from the quarry into slabs or other required form, and also for squaring up and facing blocks. No. 3 is a stone-rubbing or surfacing table. No. 4 is a stone-dressing machine. No. 5 is a combined stone-planing and moulding machine with horizontal cutter barrels. No. 6 represents a stone-moulding and planing machine with vertical cutter barrels. No. 7 is a grindstone. No. 8 is an emery wheel-cutter grinder. No. 11 is stores, and Nos. 12 and 13 offices. No. 14 underground shafting. No. 15 tramway. For repairing cutters, machines, &c., and tool-making, a small shop for blacksmith and fitter should be added; this, in addition to the usual tools, should contain a lathe, which could be driven from the main shaft.

It will be found most convenient to arrange the main shafting (No. 14), which is 3in. diameter, underground; it should make about 100 revolutions per minute. Owing to the stone-dust flying about, especial care must be taken as to the lubrication of all bearings, which should be closely fitted with sheet-iron guards to keep off as much of the dust as possible. As a lubricant we can recommend the following:—good lard oil 75 parts, powdered plumbago 25 parts. If the bearings are heavy and subject to much strain, the proportion of plumbago may be increased up to 40 parts. Should a water-wheel or turbine be used as the motive power, a reservoir should, if possible, be formed, or at any rate a considerable reserve supply of water obtained. As regards the most suitable types of water-wheels to employ, the user must, of course, be guided by the exigencies of the site, water-supply, &c.; but for information on these and other kindred points, as regards motive power, we refer our readers to our recently-published book.* Should a steam-engine be employed to give the motive power, we are in favour of a long-stroke engine, running at a moderate rate of piston speed—say an engine with a stroke twice the diameter of the cylinder. In this case the steam should be cut off early and expanded for the rest of the stroke. Where low first cost is not an object, a compound engine is to be preferred. In selecting an engine, it should be borne in mind that large cylinder diameter does not necessarily mean large power; it depends also on the mean pressure, and the piston speed. All the wearing surfaces and steam passages should be in proportion to the cylinder diameter: if this latter is too large, the cylinder pressure is reduced, and a considerable amount of steam wasted. As regards the boilers, under ordinary conditions Cornish or Lancashire boilers fitted with cross tubes are perhaps, taken altogether, to be preferred; but should the water used be very pure and the

* "Saw Mills: their Arrangement and Management." Crosby Lockwood & Co.

attendant efficient, a locomotive or tubular boiler may be used with advantage. An overhead travelling crane should traverse the full length of the mill and yard, and ample facilities in the shape of hoists, blocks, and falls, stone trucks, &c., for readily lifting the stone on and off the various machines and moving it from place to place, should be provided. The plan on which the works (Fig. 1) are designed is that the rough stone should enter the mill at the end near the engine-house and pass immediately to the horizontal and circular saws (Nos. 1 and 2); it thus becomes at once reduced in bulk and rendered more portable on its passage through the various dressing, shaping, and moulding machines. Elaborate or recessed work, &c., which requires finishing by hand after leaving the machines, passes by means of the tramway to the masons on the opposite side of the mill. If the blocks are brought from a quarry some little distance off, they can readily be hauled along the tramway into the works, either by the engine itself, or by means of a hauling apparatus, which may be fixed below the floor, or carried in hanging brackets. Should it be found convenient to fix it below the floor, it would be necessary to pass the chain or rope over the snatch block, so as to bring it on to the level of the mill floor. The apparatus consists of a cast-iron barrel driven by toothed gearing, motion being imparted by a belt working on to fast and loose pulleys, which can be thrown in or out of gear as required. In many quarries a water-balance tank on wheels and running on an incline is utilised for bringing up blocks; its utility, however, depends on the exigencies of the site, but in some cases it may be made very serviceable. The centre of the works between the tramways may be used for storing blocks of stone, so that they may readily be placed on the various machines as required. Sufficient space should be allowed between all the machines for the ready handling of the stone. The floor of the works should be kept damp and the stone-dust cleared away periodically, as it is injurious both to the workmen and machinery. If the establishment is a permanent one, the floor should either be of concrete or asphalted. A tunnel, passing in a line under the various machines, as is usual in modern wood saw-mills, can also be used with advantage in getting rid of the stone-dust. In arranging the works care should be taken that the motive power is ample to drive the various machines, as irregularity in driving, through insufficiency of power, is detrimental to quality and output.

Should the boiler-power be ample, and it be found necessary to erect an additional saw-frame, a small horizontal engine with a long stroke may be coupled directly on to the saw-frame; for a double frame, arranged to cut two stones at the same time, an engine with an 8in. cylinder by 24in. stroke will be found suitable. A fly-wheel of extra heavy section should be used to balance and secure uniformity in running.

In the stoneworks or quarry, a method of working should in all cases be decided on, and good order kept; dirt and confusion are in every way objectionable. Each man should know his own particular work, and be made to do it without hurry or bustle. All the plant and tools should be under the control of one man, who should be answerable for their condition and efficiency. We consider a system of piecework advisable wherever it can be practised.

Before deciding on the exact type of machine or cutters to employ for shaping or moulding, the stone should be thoroughly well tested, as the cutters or type of machine that may suit one kind of stone may be quite unsuited to another.

CHAPTER IV.

STONE-SAWING FRAMES.

The first operation in the conversion of stone, after it leaves the quarry, is the dividing it into blocks or slabs of suitable or convenient size for general building purposes. In doing this by hand, a blade of soft iron is used; this is usually about 4in. wide, and is fitted into a rectangular wooden frame, which is expanded or kept asunder by a pole, which rests at each end against a loose block of wood, called a bolster; the blade is generally kept in position by means of pins, which pass through either end of it, and fit into notches, formed in the side-pieces or heads of the wooden frame. The blade is strained by a chain, arranged in two parts, but connected by a rod, on which is turned a right and left-handed screw, which is fitted into nuts. Holes are provided in the rod, in which a lever or "tommy" may be inserted, to tighten up the chain, as may be required. In sawing stone the edge of the blade is usually rounded and used with a rocking motion, so as to make it bite deeply, first in one place and then in another, rather than uniformly all along the cut. We may here remark, that the term "sawing," as applied to cutting stone—with the exception of Bath and other very soft stones, for cutting which, saws with teeth are now generally used—

with a straight blade of iron, is a misnomer, as the piece of iron used is not a saw at all, its edge being smooth and unprovided with teeth; its action is really an abrading one, and not a cutting one, the particles of stone being separated or divided by friction. As, however, the term "stone-sawing" is universally employed, to make ourselves understood, we shall continue to use it. The stone is divided by the blade acting on sand and water, the small hard particles of the former taking the place of the teeth of an ordinary saw, the water at the same time somewhat softening the stone and keeping the saw-blade cool. For cutting soft stone, coarse sharp sand is used, whilst for hard stone or marble very fine sand is employed, and, preferably, that containing flinty particles. The sand should be carefully washed and sifted, so that the sandy particles do not exceed a certain size, and to clear it of extraneous matters, and the cutting-action of the saw is thus much improved; this is a point, however, often neglected or lost sight of. The fineness of the sand should be varied according to the nature or hardness of the stone to be cut, as should sand of an improper degree of fineness be employed, the work of finishing or polishing the stone is considerably increased. The quantity of sand and water necessary must be left to the experience of the workman. When sawing by hand, to overcome the excessive weight of the saw and frame, a pole or frame fitted with a cord pulley and counterpoise weight is usually suspended over the saw; this of course lessens the labour of working the saw backwards and forwards, but its cutting-speed is reduced at the same time in proportion to the weight suspended. Care must be taken in working that small stones do not get under the saw, or it will roll backwards and forwards without cutting.

In sawing stone, whether by hand or machinery, extreme care should always be observed in marking out the stone, so that it may be converted to the best advantage, especially noting any faulty places, so that the part of the stone containing them may not be entirely wasted. All lines should be carefully plumbed before commencing to saw, and very great care taken that the saw or saws run in an exact vertical plane, or "galls" will be formed, which will have to be ground away —with considerable loss of time and labour—on the rubbing-bed.

Keeping the saw in a vertical line, when cutting by hand, is a matter of great difficulty, the narrowness of the blade rendering it specially liable to twist. Taken altogether, from its slowness and the constant vibration of the saw from the unevenness of the motion given to it by the workman, stone-sawing by hand, except for small blocks, is a tedious and unsatisfactory process, and one that can well be replaced by machinery. We shall have something to say as to the guiding of saws elsewhere.

As these pages may fall into the hands of students or others unacquainted with stone conversion, we may here with advantage briefly explain the action of a horizontal stone-sawing frame worked by steam or other motive power. The machine consists of a main iron or wooden framework, in which is suspended by four weighted chains working over pulleys fitted at the top of the main frame, a rectangular iron frame longer and wider than the block of stone to be cut. In this frame are fitted, in a vertical plane, a number of wrought-iron blades or saws set at such distances apart as it is desired to cut the stone in thickness. The blades are fed into the stone by their own weight and that of the swing-frame which is a little in excess of the counterbalance weights, which are attached to the ends of the chains, by which the saw, swing, or vibrating frame is suspended. A horizontal reciprocating motion is imparted to the saw-frame by means of a slotted pendulum, to which is attached a connecting rod and crank, which receives its motion from a main or counter-shaft. The saws and frames are raised usually by a hoist or pulley blocks, and made to rest on the stone to be cut, suspended over which is a little apparatus or cistern, which supplies to each of the blades the requisite sand and water necessary for them to work. As the cutting progresses, the saws gradually pass downwards through the stone, which is thus divided into slabs or blocks.

As regards the date when machinery for sawing stone was introduced into England, some doubt exists. It is said that machinery for sawing and polishing marble was first established at the village of Ashford, near Bakewell, Derbyshire, in the year 1748, water being the motive power. Dressed granite is stated to have been used about 1730; and Macdonald and Leslie, of Aberdeen, are credited as being the first who sawed and polished granite by machinery. Gordon Hospital, erected in 1739, was partly built of dressed stones of granite, but even then sandstone was used to form the lintels and facings. It is recorded, however, that granite ornaments were many years before this turned in a lathe and polished with sand and emery.

In the year 1777, Samuel Miller, sail-maker, of Southampton, invented a circular sawing-machine, which he called a perfectly new machine, for the more expeditiously sawing all kinds of wood, stone, and ivory; and it is stated the motive power he employed was a horizontal windmill. Devices for working in stone were invented by

Sir Samuel Bentham, in 1793, and by Joseph Bramah, in 1802. Bramah claims, in his specification, which consists chiefly in improvements in wood-working machinery, certain machinery which may be worked by animal, elementary, or manual force, and which said effects are to produce straight, true, smooth, and parallel surfaces on wood, stone, and metals.

Sir George Wright, Sir James Jelf, and a Mr. Brown, early in this century, made various improvements in machinery for sawing stone. Sir George Wright's method of cutting a cylindrical pillar is worth recording, and may be thus described :—When a completely cylindrical pillar is to be cut out of one block of stone, the first thing will be to ascertain in the block the position of the axis of the cylinder; then lay the block so that such axis shall be parallel to the horizon, and let a cylindrical hole of from 1in. to 2in. diameter be bored entirely through it. Let an iron bar, whose diameter is rather less than that of this tube, be put through it, having just room to slide freely to and fro as occasion may require. Each end of this bar should terminate in a screw, on which a nut and frame may be fastened. The nut-frame should carry thin flat pieces of wood or iron, each having a slit running along its middle, nearly from one end to the other, and a screw and handle must be adapted to each slit; by these means the framework at each end of the bar may readily be so adjusted as to form equal isosceles or equilateral triangles; the iron bar will connect two corresponding angles of these triangles, the saw to be used two other corresponding angles, and another bar of iron or of wood the two remaining angles, to give sufficient strength to the whole frame. This construction will enable the workmen to place the saw at any proposed distance from the hole drilled through the middle of the block; and then, by giving an alternating motion to the saw-frame, the cylinder may at length be cut from the block, as required. This method was first described in the collection of machines approved by the Paris Academy. If it were proposed to saw a conic frustum from such a block, then let two frames of wood or iron be fixed to those parallel ends of the block which are intended to coincide with the bases of the frustum, circular grooves being previously cut in these frames to correspond with the circumferences of the two ends of the proposed frustum; the saw, being worked in these grooves, will manifestly cut the conic surface from the block.

America claims that Oliver Evans, of Philadelphia, in 1803, had a double-acting high-pressure steam-engine at work grinding plaster and sawing stone, and that he drove twelve saws in heavy frames, sawing at the rate of 100ft. of marble in twelve hours. We can find no record, however, as to the construction or working of this machine.

Some sixty years ago, a marble-sawing and polishing mill was erected in the neighbourhood of Kilkenny by a Mr. Collis; it has been represented as being remarkably simple and efficient, and is described as follows:—One water-wheel, 10ft. diam., with twelve floats, gave motion by a crank at one end of its axle to a frame containing twelve saws, and by a crank at the other end motion was given to a frame of five polishers, and beneath these again was arranged another frame carrying eight more saws. It is stated that the saws were made of soft iron, and lasted about a week; they were constantly supplied with water and sand, the latter taken from the bed of the River Nore, and washed till nothing remained but very fine and pure silicious particles. The marble taken from the mill was first polished with a stone called cove stone, which is a brown sandstone imported from Chester, and is said to be so called from being used in chimney coves. It was afterwards polished by a hone-stone, which is a piece of smooth nodule of the argillaceous iron ore, found in the hills between Kilkenny and Freshford; it received its final polish in the mills with rags and putty.

Various other small improvements were from time to time introduced, but the real father of modern stone-working machinery was the late Mr. James Tulloch, of Esher-street, Millbank, who invented, in the year 1824, and erected, a most complete plant of stone-working machinery, including straight and circular saw-frames, rubbing, recessing, planing, moulding, and polishing machines. In the patent taken out by Mr. James Tulloch, for "improved machinery for sawing stone," the machinery was arranged to work by steam or other power; and it appears from the specification that the patentee applies his mechanism in sawing and the forming of grooves, mouldings, cornices, pilasters, &c., of marble, or other stone, by means of properly indented instruments, which are to traverse the face of the stone, suspended in a suitable frame. By suspending the saw or tools in this manner the inventor considered a great advantage was gained, as they were thus kept in a perfectly horizontal line, so that the face of the stone was acted on uniformly in all its parts, and the hardest parts reduced equally with the softest. We have before us an old engraving of this stone-sawing frame, and the neatness of its general arrangement merits, we think, a short description:—The block of stone to be sawn is fixed on a truck and operated on by a number of saws fixed parallel to each other in a frame. The ends of this frame are formed on the under-side into inclined planes, which run upon two anti-friction rollers, so that when motion is given to the saws, each end of the frame will alternately be lifted up, and allow the sand and water (supplied from a small cistern) to descend into the fissure. The anti-friction rollers are attached to two slides, placed in grooves in the upright posts, and are suspended by two chains wound round drum-barrels fitted to a shaft which revolves in bearings fixed above on the main wooden framing of the machine. This shaft carries another barrel and a large grooved pulley. Over the latter, by means of a rope, is suspended a weight, which partly counterbalances the weight of the saws and frame; a chain passes round the third barrel, and is attached at the lower end to a sliding piece on a vibrating beam. Motion is given to the saw-frame by toothed gearing acting on the separate crank-shafts, which, through the medium of connecting rods and the vibrating-beam, gives alternate motion to the saws. The several pulleys to which the frame is suspended admit of its regular descent, and with a uniform pressure, as the weights of the saws and frame are heavier than the counterbalance. Taken altogether, this machine must be held to be thoroughly practical, and reflecting great credit on its designer; and whilst discussing the various other machines, we shall again and again have to refer to Tulloch's name.

In the year 1833, a Mr. G. W. Wilde took out a patent for improvements in the sawing of marble, or other stone, by means of a revolving circular metallic plate, smooth, or not serrated on the face or edge, and applied with sand and water, as is done with the straight saw; and also for marking on the surface or periphery of a metallic or wooden cylinder or wheel, the converse of the intended moulding or grooving, by means of which a series of mouldings or grooves can be wrought on a surface of marble or stone at one operation, with sand

and water, and in like manner polish with putty, buff, or pumice-stone, or other polishing material.

In 1843, Mr. J. C. Wollaston took out a patent for improvements in machinery for cutting marble and stone, which included special appliances for cutting pipes or tubes of those materials. About the same time (1843), a stone saw-guide was patented by Hutchison; this consisted, briefly, in making the saw-frame run between guides of wood, which extended across the room when the saws were at work.

In a horizontal stone-sawing frame of recent construction a new form of skeleton frame has been employed. The sawing or swing-frame is made without sides, having only a front beam, and a peculiarly formed back beam. The sawing plates are tightened by means of rods or plates connected to ears or lugs on each of the upper portions of the back and front beams. Through the centre of the beams are shafts working in bearings fitted in adjustable boxes carried by roller carriages. The roller carriages have each one or more rollers, or antifriction bowls, to enable them to work freely up and down the face of the upright pillar or main framing of the machine. Each roller carriage has a projection fitted with steps to carry the beam shafts, and each roller carriage is carried by chain or rope from its own end. The chain or rope passes up the face of the pillar over a pulley on the top beam, around a shaft or spiked wheel; then along the top beam over another pulley, and down the face of the other upright pillar, and attached to the other end of the roller carriage, where it is fixed. The sawing-plates are cottered and kept perfectly tight by the resistance of the rods or plates connecting the ears or lugs of the end beams of the saw-frame.

The author has recently tried with success sawing

stone and marble by means of an endless steel band, after the fashion of a band saw for sawing wood, but in lieu of teeth, the edge of the blade was kept rough, and supplied with sand and water or emery in the usual way. For irregular sawing this will be found extremely useful; the chief difficulty to contend against is the saw buckling and running from the line. Speed of saw about 250ft. per minute.

The main framing of stone-sawing machines is preferably made of iron, as any little extra outlay thus incurred is counterbalanced by increased stability and freedom from vibration. With this end in view, it will be found advantageous to arrange the trucks carrying the stone to be sawn to run on the foundation frame, so that the whole weight of the stone and the trucks is concentrated on the machine, and utilized in overcoming the vibration caused by the saws when in motion.

Should the framework be constructed of timber, it should be very firmly braced together, and bound with iron at all the joints, care being taken that the timber employed is sound and well seasoned. The vibrating or swing-frame which carries the saw should combine great strength without excessive weight, and with adequate arrangements for setting the blades. The chains by which the swing-frame is suspended to the cross shafts of the main frame should work freely in pulleys, and a selection of weights supplied, so that the frame may be accurately balanced, and adjusted according to the nature of the stone or the speed it is desired to cut. This is a point of importance, though often neglected if a series of weights are not to hand. A good sawyer, on seeing the nature of the stone he is operating upon, should adjust his weights, or in other words, the speed of his cut, to the greatest nicety. Means should be taken that the swing-frame can be readily raised and lowered, and that the cutting-blades are kept in an exact vertical plane. The pendulum used to give motion to the swing-frame should be slotted for a considerable distance, so as to allow the saws to pass through large pieces of stone. The swing-frame must have a steady and uniform stroke from top to bottom of the cut. Where more than one machine is employed, it is preferable to drive them by means of cranks and connecting-rods, on to a countershaft fitted with fast and loose pulleys, as thus one frame may be thrown out of work without stopping both, as is the case when the connecting-rod is coupled directly on to the main shaft. Adequate means must of course be taken to afford the saws a constant supply of sand and water. A complete set of packing-pieces or other means of adjusting the saws to equi-distances apart should in every case be to hand.

Various plans for supplying sand and water to the blades are in use. The following can be recommended on account of its simplicity and efficiency:—Make an iron cistern for water about 2ft. 6in. long, by 1in. high by 1ft. wide, and fit on each side of it, say, 15in. water-cock. The water will be allowed to flow slowly through these cocks and fall into an equal number of grooves formed in the bottom of boxes filled with sand. These grooves should be so constructed that the sand and water flow out together, through suitable openings on one side of the boxes, and directed into the various saw fissures. A grooved sand-box should be arranged on either side of the cistern, so that two streams of sand and water are directed to each blade, one on either side of it.

For vertical guides to the saw-frame we prefer anti-friction rollers fitted in small carriages, and arranged to travel freely up and down the face of the vertical pillars of the main framing; this plan will be found to work with much greater freedom than the ordinary sliding arrangement generally employed.

Where it is required to saw and polish marble, &c., it will be found convenient to arrange a double-ended machine or mill, a reciprocating motion being given to the saw-frame and polisher at the same time by connecting rods and cranks attached to a main or intermediate driving-shaft. The saws, or more properly speaking, cutting instruments, are made of wrought iron plates, and are similar to those used in sawing ordinary building-stones, and the general arrangement of the saw-frame is the same. The polisher at the opposite end of the machine is arranged with a longitudinal motion, and the slabs of marble, after being sawn from the block, are one by one placed on a travelling carriage, mounted on rollers, and running on rails, and to which a lateral reciprocating motion is given. The polisher is arranged to bear on the surface of the marble, and from the lateral motion given to the travelling carriage, is enabled to operate on every part of its face. A simple and efficient mode of obtaining the lateral motion is to arrange on a crank-shaft passing beneath the travelling carriage a worm and worm-wheel, and by means of another adjustable crank, motion can be given to the worm-wheel, but care must be taken that the adjustable crank is at all times longer than the crank on the worm-wheel spindle; by this plan the wheel will vibrate through an arc, and not make an entire revolution; thus by using an additional toothed wheel, and by attaching a rack to the travelling carriage, a lateral reciprocating motion can be obtained, which can be varied in the length of its traverse by increasing or reducing the throw of the adjustable crank before mentioned. The polisher is usually put in motion by an upright vibrating frame, which vibrates with the stone saw frame at the other end of the machine, and receives its motion from the same crank.

The main framings of all machines used for sawing granite, marble, and other very hard stones, should be made of iron instead of wood; they should be of heavy section, and designed to overcome excessive vibration in working.

For sawing marble slabs into strips, either a ripping bed, carrying a number of discs, or a saw worked by hand, called a grub-saw, is generally used. The grub-saw is an iron blade notched at the edge, and stiffened by a backing of wood, like the metallic back of a tenon-saw; the kerf for the saw is usually started by means of a narrow chisel. For cutting marble into cylinders, a rod or wire saw, of triangular section, is often employed, the marble being mounted on centres, and revolved against the saw.

Taper slabs may be sawn by means of two sets of saws; these may, however, be worked with a crank, by connecting the two sets of crossbars by means of links, so that the adjustment of one determines that of the other, and by connecting the crank to the first set, motion is imparted to both. The saws may be made to cut taper by adjusting one set laterally, by means of the crossbars, to any desired angle with the other, the movement of the crossbars corresponding to the relative inclinations of the guides; these may be kept in any required position by means of bolts.

The most rapid cutting sand for general sawing is that containing a large amount of flinty particles, but the difficulty of obtaining this in some districts has induced the author to construct a machine capable of breaking and crushing 8 tons of flints per day, to a size that will

FIG. 2, 3.—HORIZONTAL STONE-SAWING MACHINE.

pass through a sieve either 10, 12, 14, or 16 gauge to the inch as may be required.

Fig. 2 represents a horizontal stone, granite, or marble sawing-machine, with engine attached, from the designs of Messrs. Rushworth Bros. The main frame of the machine is constructed chiefly of steel and iron strongly braced. The swing or saw frame is made of mild steel straightened by diagonal strips, which form lattice work, thus combining, as far as may be, strength with lightness. The swing frame is raised and lowered by means of bevel and spur gearing, working two square threaded screws fitted with gun-metal bushed nuts; it is suspended from these nuts by four connecting rods fitted with adjustable coupling boxes for balling the frame. The nuts are fitted to work down the main upright pillars of the frame. The reciprocating motion is given to the frame from the counter-shaft by means of a crank and connecting rod working down the centre of the lattice trunk. The different motions of this frame are all automatic, and the designers claim that, by an arrangement of cone pulleys, they are enabled to readily regulate the cutting speed of the frame, according to the nature of the stone being sawn, without the machine being stopped; and, from its peculiar construction, that it can be run at a much higher speed than is usual with machines of this class. It is claimed that under ordinary conditions these frames are capable of cutting at the following speeds : Portland stone, 21in. in depth per hour; Yorkshire, 15 to 18in.; Grit, 12in.; Rossendale, 12in.; Blue rock, 12 to 15in.; Granite, 6in.

CHAPTER V

STONE-SAWING, &C. WITH DIAMOND POINTS.

STONE saws, fitted with diamond or "carbon" points, have been used with some little success, but chiefly in America. They have usually been applied to circular saws; but reciprocating jigger and band-saws have also been mounted in this way. One of the difficulties found in working carbon-pointed saws is the fastening the points securely in their seats, and a number of patents with this end in view have been taken out. Some of these hold the diamond by fingers clamped in sockets, or embed it in the saw by means of sockets or solder, or hold it by clamps kept in position by wedges. This difficulty being surmounted, we think the ordinary band-sawing machine, so equipped and running at a slow speed—say, the teeth or points to travel about 250ft. per minute—should prove a valuable tool for curved stone-sawing, such as that required in Gothic windows, arches, &c.; but for rough, heavy block sawing—bearing in mind the increased first cost, and the difficulty of keeping the carbons in their seats—we fail to recognise at present the value of diamond saws for this purpose. The diamonds usually employed for sawing are black, and are chiefly found in Brazil. Like all "carbons" they are of extreme hardness, and will, when properly held, divide granite, marble, and almost any other mineral, the difficulty, as we have before remarked, being to hold them; and this difficulty, practically speaking, has been admitted by an American exhibitor in the Philadelphia International Exhibition, 1867, as he attached to his machine a sieve to catch the diamonds when they were dragged from their seats, and prevent them being washed away. The sockets carrying the diamonds are necessarily wider than the saw-plate, but by "setting" them to the right and left-hand, in a similar manner to a saw for wood, this difficulty is got over.

Dressing hard stone by means of diamonds has also been attempted, and large sums of money have been spent thereon; but with no practical commercial success. Diamonds have been used for boring hard rock, granite, &c., for blasting or sinking, with very considerable success; but the action of boring, it must not be forgotten, is entirely different from that of sawing, the tearing or pulling action on the diamonds being absent. Mr. J. T. Gilmore, of Ohio, in 1863, patented a system of dressing millstones, fluting columns, dressing building-stones, and making mouldings, by means of diamond points mounted on a revolving disc; a considerable number of other patents in connection with stone-working by means of diamonds have been taken by Young, Emerson, Husband, Gear, Dickinson, and others.

Dickinson constructed a variety of carbon tool-points for dressing and working stone; for turning stone he used a triangular prism-like cutter; also for various other stone-working operations he designed carbon points in the shape of hexahedrons, double-sided trapezoids, drill-faced parallelograms, truncated prisms, quadrangular double-faced points, quadrangular pyramids, flat octahedrons, flat ovoids, tetrahedrons, &c., &c.; but notwithstanding the large amount of ingenuity and money expended in trying to develop stone-working by means of diamond points, we are afraid that, with the exception of drilling, turning, and perhaps grind-stone dressing, at the present the disadvantages attending the use of diamonds more than counterbalance any supposed advantages.

The invention of the diamond drill is ascribed to Hermann, who patented it in the year 1854, and claimed as novel the use of crystals, or the angular fragments of thick diamonds embedded by alloys in a metallic stock, for working, drilling, and turning hard stones such as granite, porphyry, marbles, &c. For drilling purposes he inserted the diamonds in holes drilled for them in the end of the drill-rod, the metal being battered down around them to form a bevel. The drill was arranged to slide vertically, and was rotated by bevel gearing; water was used when drilling. Leschot, from the year 1860 to 1864, designed various forms of drills, and Pihet, in 1866, introduced the annular drill-head, which is a steel ring studded with diamonds; these, we believe, have been used for sinking and blasting purposes at the Mont Cenis Tunnel and other large works with success.

The carbon-points are set in such a manner in the ring or cylinder of steel that they project alternately on both the inside and outside of the periphery, and thus cut a clearance in the rock for the drill-bar. The drill-bars are usually made hollow, and the core of rock cut out by the drill passes up through the bar. Water under pressure is supplied to the drill to keep it cool, and wash away the rocky particles as cut.

For cutting grooves or channels in marble, a chain-saw with its links mounted with diamonds has been used; the diamonds were held in the ends of split screw bolts or clamps, which were tapered and forced into tapering holes, causing them to contract upon the diamond. The diamond chain-saw was made to run round two wheels, with arrangements for feeding it to the stone, and for reversing its cutting-action. Diamond points have also been used for recessing and mou ding stone. These were fixed in a circular tool, mounted on a vertical revolving spindle, after the manner of a recessing machine for working wood. The overhanging frame was arranged to pivot in any direction, and the stone to be worked was placed beneath the cutting-stool. The difficulty of safely holding the diamond points being overcome, this form of tool should be very useful in recessing and countersinking hard stones.

CHAPTER VI.

CIRCULAR SAWS FOR CUTTING STONE.

THE earliest form of circular sawing machine used in stone conversion was the old ripping bed; this was probably used for the first time in this country about the commencement of this century. During the last twenty years, machines carrying circular saws of large diameter for sawing, facing, and edging stone, have been introduced. These saws are usually made of wrought iron or steel, with adjustable or false teeth fitted into their periphery. Circular saws waste a considerable amount of stone, and take more power to drive than frame saws; but are very expeditious, as from 150 to 250 running feet may be cut in a day of 10 hours in a stone of moderate difficulty. The saws are arranged either vertically or horizontally, and the shape of the saw-teeth or cutters should be varied according to the nature of the stone being worked; but these points, owing to the varying nature of stone, can only be correctly determined by practically experimenting on the stone itself. For edging and squaring up stone, and to save hand labour in jointing, two or more saws are mounted on the same spindle, and set so far apart that they just cut the edge of the stone on either side, and make it parallel; it is then turned round, and set by the square side, and the other edges are treated in a similar manner. The general arrangement of these machines is usually somewhat similar to that of a planing machine for iron; but in lieu of the ordinary cutter-box and slide, a horizontal or vertical saw spindle is mounted, the stone to be sawn being placed on a travelling table which traverses immediately beneath the saws. This table, if of cast iron, should be of substantial construction, or it may be liable to fracture if a heavy stone is suddenly placed on it; this is especially the case in cold weather. To overcome this, tables are occasionally made of oak, firmly jointed together, and mounted on iron plates. If timber is used, we can recommend its being faced with wrought-iron plate, as should the stone be placed directly on the wood, it will rapidly become rough and uneven, and difficult to move the stone about on. For general builders' work the saw-spindle should be made to rise and fall, so as to adapt it for checking and other similar purposes.

If the travelling table is mounted to run in V-slides, we can recommend for lubricating purposes the formation of recesses at intervals of 6ft., in which an iron plate of the same shape as the slide and covered with felt can be held in position by a spiral spring, and the recess being supplied with oil, it is lifted against the V of the sliding-table, and thus keeps it constantly lubricated, and with much less waste than is occasioned by the oil being carried along the whole length of the slide, as is the case in most of the machines hitherto made. The travelling table should be traversed by means of a screw, as it will be found steadier than a rack and pinion, and we prefer the table mounted on a dovetail slide, as it is less liable to jump after each traverse of the cutters or teeth.

The old ripping bed consists of a kind of bench or table, in which a number of circular saws are mounted on a rising and falling spindle, by which they can be regulated to the depth of the cut. The marble, slate, or stone to be sawn is fixed on a travelling table, which is arranged to run on rails fitted to the main framing of the bench. This table is fed slowly forward by a screw or counterpoise weight and rope. The saws are discs of wrought iron without teeth, and are fed with sand and water, in similar manner to straight saws. They are divided on the spindle by a series of washers, which can be regulated according to the width of the slabs to be cut. A "feather" extends the full length of the saw-spindle, and prevents the saws turning round. The saws are usually arranged to cut upwards.

Stone saws should always be arranged to rise and fall, and also have lateral adjustments, so that they may be readily set to suit any depth or width of slab. Care should also be taken that the stone is securely "dogged" or held in position, as, should it be allowed to move in the cut the stone will be "galled." In the case of large circular saws, the stone should be held especially firm when the saw enters the cut. If the saw should be used for edging purposes, and a very thin cut taken, it should be supported and guided laterally to keep it from running out of the cut. In working stone, it is of the utmost importance that the cutting-tools employed should be shaped and tempered to suit the different kinds of stone. For example, some stone may be cut or forced off in large chips, whilst others have to be scraped away. Soft or rotten stones are better cut with straight than circular saws, and by well weighting the swing-frame they can be sawn with tolerable rapidity.

The most important point in connection with circular saws for cutting stone are the teeth; these should be of simple form, easily made, and readily removed for sharpening. Owing to the constant grinding action of the stone, and the consequent friction on the teeth, these are invariably made renewable or "false," and the cutting is performed by them, instead of by teeth formed in the periphery of a plate of steel, as in sawing wood. These false teeth undoubtedly waste more stone than an ordinary saw would; but this is more than counterbalanced by the fact that by their use the saw always remains the same diameter, whilst the ordinary plate would be rapidly worn away and rendered useless.

The invention of movable tools or teeth for sawing and working stone is due to Mr. George Hunter, who, in conjunction with the late Sir W. Fothergill Cooke, made many improvements in machinery for the conversion of stone: in fact, they must be considered, without doubt, to be the pioneers of the more advanced type of stone-working machinery. These improvements extended over a number of years, commencing in the year 1865, and they claim the invention of movable tools or teeth capable of application to machinery for a variety of purposes: firstly, to the sawing of blocks and slabs of rock of considerable thickness for building or other purposes; secondly, for facing the surface of squared-up stones in an ornamental manner; thirdly, for undercutting stone, slate, or coal in situ, when the rock lies more or less on the incline, and also for the vertical cutting of the living rock.* These machines were manufactured by the late firm of Powis, James, & Co., under the superintendence of the Author, and very excellent work was produced by them, in Bath, Portland, York, and other stones.

Fig. 4 represents the movable cutting-tool invented by Hunter, fixed in its socket or holder. It consists of a bolt made of the best rod steel; the head is forged into a

---

* See Journal of Society of Arts, Vol. XV., p. 19.

cupped or trumpet form, turned at the edge and then hardened. When in use it is simply slipped into its socket, which is also made of steel. When the edge is

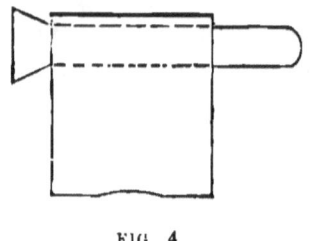

FIG. 4.

dulled or chipped, the tool is turned in its socket so as to offer a fresh cutting margin, and as it wears away on the advancing side, the tool will offer several fresh cutting surfaces before it is entirely worn out. These tools are varied in size according to the circumference of the sawplate, and range from 4in. to 8in. long, and the cutting head itself from ½in. to 1¼in. wide. The length of the tool allows of the head being softened, again set up, turned, and hardened, until it is too short for further use. The stems of the sockets are of the same thickness as the saw disc that receives them, and are slipped into grooved openings made in the periphery of the disc.

The other tools for stone-dressing, used by Messrs. Hunter and Cooke, are illustrated in Figs. 5 and 6, and,

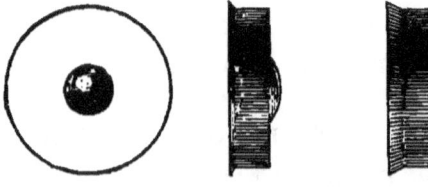

FIG. 5.

as in stone conversion, the cutting tools are, without doubt, the most important factors of successful working, we shall briefly describe them.

Fig. 5 represents a tool which consists of a disc of steel punched from a plate, and shaped precisely similar to the head of the bolt-like tool just described. It is ready cupped and sharpened, and made with a boss behind to fit into a corresponding groove in the back of its holder, and with a hollow in front to receive a set-screw. These discs can be punched into the exact form required, and only need tempering to be fit for use. The holder of the disc-tool so grasps it that its cutting-rim projects only very slightly beyond the holder; it thus offers very little leverage to the resistance of the stone, and therefore rarely becomes loose. Fig. 6 represents another form of cutting-tool: it is formed out of a symmetrical, acute-angled trapezium, cut from a ribbon of steel, two or three-six-teenths of an inch thick, according to the roughness of the work it has to do. To convert these strips of steel into the required form of tool, they are bent sharply on their middle, so as to bring the acute angles opposite to each other, but slightly turned out of their cutting-angles.

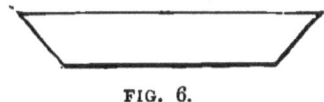

FIG. 6.

These tools are cut from ribbons of steel, and a number of the flat slips are screwed up together in the vice and sharpened; they are then bent and hardened, and are fit for use. The socket for this form of tool is extremely simple (Fig. 7); it is merely a hole into which the tool slips, and in which it is held by the spring of its own arms. It was designed for roughing down the face of the stone, preparatory to the use of the facing-tool. Messrs. Hunter and Cooke also employed flat, concave, and other forms of tools for giving ornamental surfaces to ashlars, quoins, sills, &c.

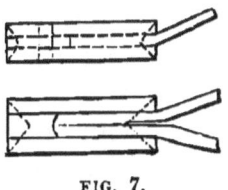

FIG. 7.

In the design of their early circular sawing-machine, which worked these cutters, there was no especial novelty; it consisted of a table moving on V-grooves, or on friction pulleys, to which the stone to be cut was fixed by cramps, or when very massive and rough by chairs. The table was fed forward by a screw, at a speed varying from three to six inches per minute, the speed being varied according to the nature of the work being done. The saws were mounted in collars, which could be adjusted laterally, and from one to four were used at a time. They also claimed the use of a succession of travelling tables for carrying the stone, which were always advancing, and on which blocks were prepared before the saws were ready to receive them. This was with the idea of saving the time of running back, unloading, and preparing another block; the table was then either to be sent forward with a fresh load to another saw, or lifted by a crane on to a line of rails parallel to the series of saws, and run back to commence its course again. There are, however, several practical objections to this plan, and we have never heard of its being brought largely into use.

A further application of Messrs. Hunter and Cooke's movable cutting tools was for dressing and facing stone. To perform this, the saws were removed from the horizontal spindle, and in their place a removable cylinder was slipped on or bolted on the spindle in halves, and this received the cutting tools. These were fixed in toolholders, placed spirally round the cylinder. The object of the inventors in arranging the tools spirally was to obtain a divided and regular pressure upon the face of the stone at intervals of 2in. or 3in., always nearly uniform, but ever passing spirally from right to left, this plan doing away with the objection of a large number of cutters striking the stone at the same time, as it will readily be seen that with tools arranged spirally only a few of them would be acting on the stone at the same time the remainder entering and leaving the cut at the same moment. As many of the latest forms of machines and cutting tools for working stone have been based on Messrs. Hunter and Cooke's patents, including the Ridge stone machine for cutting ridges out of each other, thus:

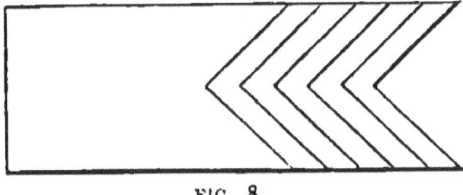

FIG. 8.

—we think they are fully entitled, as we have before remarked, to be considered the forerunners of modern stone-working by machinery. As regards the earlier constructors of Stone-working Machinery of which we have any reliable records, the names of Mr. James Tulloch of London and Mr. James Hunter of Forfarshire will always hold a foremost place.

In practice circular saws are occasionally mounted on vertical spindles, the collars on which the saws are fixed being fitted with a feather, which works in a corresponding groove on the spindles, thus allowing the adjustment of the saws up or down to suit the work. The stone is cramped on a travelling table, and fed forward in the usual manner by means of a screw or rack-feed ; but we think saws arranged on a horizontal spindle will be found generally more handy and easy of adjustment.

The objection urged against cutting stone with circular saws is the cost of the cutters or teeth; this objection in some cases was tenable, and has gradually led to the abandonment of the more complicated forms of cutters, which may possibly theoretically have been of the correct form ; but their first cost, constant renewal, and difficulty of keeping in order has led to the adoption of the simplest forms, at a considerable saving, and we may here observe the cost of tools for cutting, say, 200ft. run of sandstone should not exceed 5s.

Some years ago circular saws with solid teeth were tried with some success, but the cost of renewals was found to be excessive; the saw-plates were ¼in. thick, and there was ⅛in. set on each side of the teeth, thus giving a cut ½in. wide; the speed of the saw at the periphery was about 108ft. per minute.

Whatever teeth are employed in circular stone saws, they must, to be commercially successful, be of the simplest possible form, easy to make and renew. Our illustration, Fig. 9, represents saw and facing-disc cutters

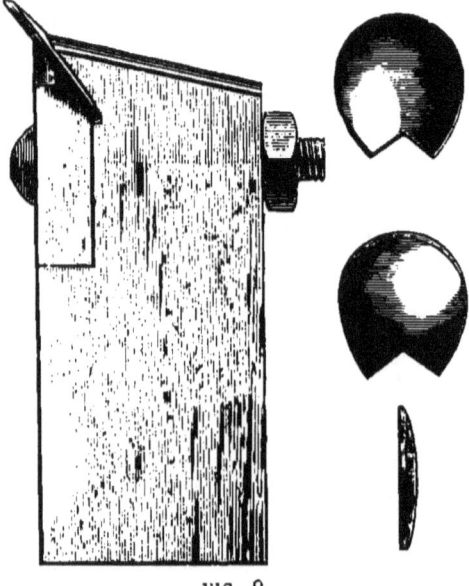

FIG. 9.

(Stevenson, Rea, and Dunlop's patent), and also cutter-holder or socket. An advantage claimed for these cutters is that the sharp edge they have when new is retained till they are worn out; they certainly also have the advantage of simplicity, and are apparently based on one of Hunter's cutters, Fig. 5.

For cutting window-sills, coping-stones, steps, &c., a number of saws can be mounted on the same spindle, and five or six cuts taken through a block at the same time, and thus also a large amount of stone often thrown on one side may be converted into useful and saleable articles. Unless the blocks are large the saws need not exceed, say, 4ft. diam. These are arranged to cut uphill, and the stone must be securely cramped on the table. When the saw spindle is placed below the travelling table, the saw must be arranged to cut downwards, or the saw teeth will strike the stone abruptly with a direct jarring blow, which will in many cases twist and buckle the saw-plate and strip the teeth.

As regards the cutting-speed of circular saws for stone, no arbitrary rule can be laid down, as this must depend on the nature of the stone operated on. A speed at the periphery of from 50 to 200ft. per minute, or with a cutting-speed varying from 3in. per minute in hard stone up to 12in. in soft stone, will be found suitable.

In cutting very difficult stone, such as that containing pyrites, the cutting-tools should run very slowly indeed, say 40ft. per minute, or they will be found to heat red-hot, and will, of course, at once be rendered useless. The feed should not exceed 2in. per minute. Some difference of opinion exists as to the advisability of sawing stone with circular saws, with or without water. Some stones can, without doubt, be readily sawn dry, but from our experience we prefer wet sawing, as it keeps the tools cool and prevents unnecessary dust.

In machines for sawing heavy blocks, the saw should be raised and lowered by means of self-acting gearing, as raising by hand is a slow and laborious process. All bearings should, as far as possible, be protected from dust. The discs carrying the teeth or cutting-tools are usually made of wrought iron, but should by preference be made of steel, which, although somewhat more difficult to fit up, are more rigid in work, and less likely to buckle, and they can also be made of a slightly thinner gauge. If steel is used, it should be of mild temper.

In contrasting machines for circular sawing and planing stone, with those for wood, it will be found the principles of operation involved are directly opposite. In working wood, the action is either a splitting one—with saws—or a cutting one, with plane and other irons; the number of revolutions of the tools is also very great. In the case of stone-working, the action with fixed tools in which there is a dead contact with the stone, is essentially a grinding one, and with revolving cutters the stone is "spauled" or levered off; the speed of working is also slow, the pressure, however, on the cutting tools and bearings is usually much greater than that of wood. With circular saws for cutting stone this is especially the case, the spindle carrying the saws should therefore be strongly supported by massive side standards, and have ample bearing surfaces; the whole framework of the machine also should be of massive construction, to overcome excessive vibration in working. Should there be a jar on the saw cutters or teeth in working, they will be found to deteriorate much more rapidly, and the work turned out will not be so true on the face.

With a well constructed circular saw the stone should leave the machine sufficiently true on the face that it may be bedded or jointed without further preparation, either by hand, or on the rubbing bed or planing machine. For rapidly squaring large blocks of stone for harbour and similar works, circular saws will be found especially valuable, and, as they become better known, their use should be largely extended. For dividing very large blocks, two circular saws, placed one above the other, but working in the same vertical line, may be used.

For joining flat stones, such as paving, all hand labour may be saved by mounting two saws so as to trim two edges and make them parallel, and by reversing the stone and setting it square by these sides, the other edges may be served in a similar way.

Circular saws for cutting stone have been constructed of as large a diameter as 13ft., and two of these were erected by Mr. George Hunter some years since for the Tyne Navigation Commissioners. The following are some of the dimensions and weights :—Saw shaft, $15\frac{1}{2}$in. diameter, weight, 3 tons; driving wheel, 3 tons, standards 3 tons each, saw plates with collars, 3 tons 2 cwt.; height above floor, 17ft.; cut, 5ft. 6in.; width of cut, $1\frac{3}{4}$in.; speed of tools, 18ft. per minute; feed of table for cut, 3in. per minute.

CHAPTER VII.

STONE-DRESSING AND PLANING MACHINES.

In designing machines for dressing stone to a plane surface, the first idea that appears to have struck the greater majority of inventors was to imitate the action of the mason's chisel or quarry axe by mechanical means, and the whole of these have ended either in absolute failure or in the very smallest degree of success, and the reasons for this are not very far to seek. To make a "steam stone-mason" commercially successful, it must be able to turn out a large amount of work, and to do this necessitates the use of a considerable number of masons' chisels. To successfully work these mechanically, may not appear very difficult in theory, but in practice we invariably find, where a large number of tools are employed, they vary in wear from difference in temper, material, or the work they have to do; and this is, without doubt, one of the great reasons of the failure of this class of machines, whether for working stone, wood, or other materials. There are, however, other reasons why the action of the mason's chisel has not been imitated successfully; it is in a degree elastic, and can be varied in strength and angle at the will of the operator, who can at the same time pick out the weaker points of the stone, and by attacking it here throw off larger chips than he otherwise could do. On the other hand, a blow given by mechanical means is a positive one, and although it may be made elastic by a spiral spring or other means, its strength or the inclination of the cutting tool cannot be varied at will, as is the case when worked by hand. Again, when a number of chisels are employed mechanically, there arises the difficulty, not only of their variation in wear, but of keeping them sharp, necessitating (even if made reversible) constant stoppage of the machinery; but the mason's quarry-axe or chisel is kept constantly sharpened by his turning it about and changing his hand, or reversing the direction of his work.

Revolving tools of various types, having a circular movement in a parallel plane to the face of the stone, are employed for stone-dressing; and vertical cutting blades acting as a knife, and toothed-blades, each tooth acting as a chisel, have also been used. We are strongly against high-speeded cutters for dressing stone, unless they can be managed with a rolling contact, in which case the attrition is much reduced; and we think it may be taken as an axiom that in all tools used for dressing stone those which produce the largest chips without "plucking" the stone, or make the least dust, are to be preferred, as the tool which produces much dust is rapidly grinding itself away, from its dead contact with the stone. A large number of cutters should in all cases be avoided, owing, as we have before remarked, to the practical impossibility of keeping the temper and consequent wear of them all alike. It is important whatever system is employed that the surface of the stone is not crushed or bruised in the working, or when fixed it will be found to much more rapidly deteriorate.

As regards the materials used for tools for stone-dressing, chilled cast iron and steel are those almost entirely employed, but there remains ample scope for an invention in this direction. What is wanted is a material combining hardness and toughness in the greatest possible degree, to withstand the enormous strain and abrasion to which the tools are subjected.

Various other plans for planing and dressing stone by mechanical means have been tried; in most of these the stone is dressed by tools fixed in a stationary tool-box or holder, whilst a travelling table on which the stone is fixed passes beneath them, after the manner of an ordinary planing machine for working iron. One system consists in arranging a series of cutters on a hollow revolving cylinder, which is provided with small perforations or openings near each cutter, and also with a stuffing-box and pipe, through which water may be conducted to the stone at the point where the tool is acting. In another plan cutters are attached to a rotating-wheel, and act upon the face of the stone, which is fixed on a travelling table, traversed by rack and pinion gear. A dressing-wheel has also been worked—but, we believe, with only small success—with slits across its periphery, and arranged to hold sand, which acted in lieu of steel or iron cutters upon stones fixed on a table which traversed beneath.

We believe that the first machine made for dressing stone in this country was that patented by Mr. James Hunter, of Leyonide, Forfarshire, about the year 1834; it was generally known as the Forfarshire stone-planing machine, and no improvement in the principle of working in that class of machine has since taken place, although the framing and general details have been strengthened. Roughing and facing tools were employed, and these were so arranged that whilst at work the attendant could adjust them to any desired cut, and could bevel or set to any cutting-angle the facing tools. For dressing Yorkshire flags a very ingenious arrangement for throwing over the tools after each stroke was employed. We believe that tertiary limestone was also successfully worked by this machine, some samples of which, although rather hard, are not much more difficult to work than the soft stones commonly employed for building purposes.

There was a stone-planing machine exhibited in the Industrial Exhibition, New York, 1858, adapted for dressing hard stone. This machine consisted of an upright frame, in which revolved a vertical shaft, carrying three horizontal arms. At the extremity of these arms were fixed circular cutters, inclined outwards about 45° from the perpendicular, or about the angle at which the workman would hold his chisel. They were about 10in. in diameter, and $\frac{3}{4}$in. thick, made of steel and bevelled on both sides, leaving a sharp edge. They were fitted upon axles, and were at liberty to revolve loosely in their bearings as their edges struck the stone. The cutters were carried round by the shaft at the rate of about 80 revs. per minute when planing freestone, and 60 when planing granite. The stone was moved forward on a bed to which it was keyed; the cutters struck its surface obliquely as they were carried round on the revolving arms, turning at the same time on their own axles, and chipping and breaking off the projecting portions at every cut. The machine is reported by Whitworth to have planed the face of a stone 4ft. long by 2ft. wide, in seven minutes.

Another modification of this machine, which was not so economical, was employed when it was necessary that the face of the stone be left in lines as it came from the tool. The stone was keyed on a travelling bed and passed under a frame, in which worked a sliding-carriage, driven by a crank; in this carriage was fixed the circular cutter at the required angle, and as the stone was carried along, the cutter was driven backwards and forwards across its face at right angles to the direction in which it moved, and chipped off parallel breadths of stone at every cut. The cutters could be used for planing from 300 to 400 square feet of freestone surfaces, and about 150 square feet of granite, without being ground.

This latter statement is given by Sir Joseph Whitworth, in his report of the Exhibition; but if it could be borne out by facts, we are at a loss to understand how it is machines giving apparently such favourable results have not been largely manufactured and introduced. As regards dressing the granite, this is especially incomprehensible, as up to the present date we are not aware of any commercially successful machine for this purpose, although much money has been spent, and many improvements in the principles of construction and manner of working have been introduced.

The Earl of Caithness, many years ago, invented a stone-dressing machine, which has chiefly been used in dressing the Caithness flags. In this machine, the cutters are arranged to strike the surface of the stone vertically; and the apparatus consists of a set of vertical, parallel bars of metal arranged in suitable guides in a substantial framing, and furnished at their lower ends with steel or hardened metal cutting or reducing edges. These bars are actuated by cams, which elevate them to a certain predetermined height, when they are allowed to drop on the face of the stone or other substance under treatment, and thus chip or cut away the surface, the stone at the same time being fixed on a bed, and made to traverse slowly beneath the cutter.

The framing of the machine resembles that of a planing-machine for iron. It consists of two cast-iron standards, carried upon a timber base, each standard having a front bracket carrying a bearing for the cam-shaft which lifts the cutter-bars. These bars, which are all guided vertically in the framing, are shod at their lower cutting ends with serrated or notched steel faces for chipping the stone. Each of them carries an adjustable stud-arm, against the under-side of which the revolving cams of the shaft strike in working. The cam-shaft is cast with all its cams—one for each cutting-bar —solid upon it. It is driven by a belt and pulleys, and as it revolves, the cams, being disposed helically upon it, lift up and let fall the cutters in succession. This gives a dressing or chipping action right across the face of the stone beneath, and produces the finished surface necessary in footways. The stone is carried by rollers mounted on a travelling carriage, which is moved forward after each revolution of the cam shaft, to present a new surface to the cutting action. In some instances it may be necessary to elevate the whole of the cutters from the stone at the instant the latter is traversed forward; this is accomplished in the present machine by a cam. On the end of the main shaft the feed-motion of the stone is affected by a link-rod worked by a stud-pin from the main shaft, the lower end of this rod being connected to the racket action of a central traversing chain pulley shaft beneath the stone. The notches of all the cutting faces except the two outside ones run in the line of the feed-motion; the two outside cutters have their notches running the transverse way to avoid injury to the arrises of the stone.

The cutters in a machine invented by Mr. Jos. E. Holmes, of Mold, about 15 years ago, were arranged to act horizontally, the power being applied by a driving-belt direct from the engine to a rocking crank driving-shaft, to which the cutters are attached. At the opposite end of this driving-shaft was an arrangement of bevelled gearing, whereby the cutters were put into or taken out of gear, and by which also the power was transmitted to a vertical spindle, which by a worm wheel operated on the travelling table. In working a block of stone down to a rough surface, two sets of cutters were employed, and so adjusted that when the motion of the table was reversed, the one set cut away the ridges left between the groves cut by the other set. For granite, black marble, and the harder kinds of stone, grooving points, rather than chisels, were used for the rough cutting down to a uniform level. The finishing tooling was done by a broad chisel working across the entire face of the stone. The accompanying sketch (Fig. 10) will

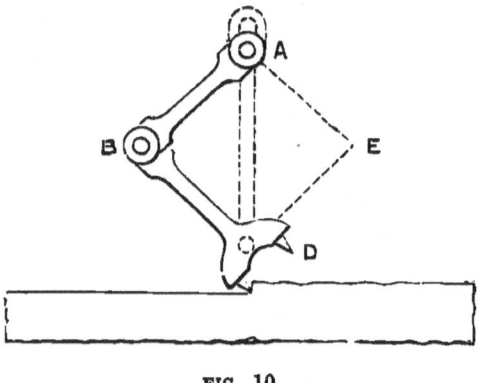

FIG. 10.

convey an idea of the form and action of the working arm and cutters. The rocking shaft at A gives the scooping action to the range of cutters. When the stone has passed for its whole length under the cutters, the motion of the table is reversed, the pin at B is taken out, and the arms are joined again at E, and the cutters or dressing tool, as the case may be, at D are brought into action.

It is evidently an attempt to imitate by mechanical means the elbow-and-wrist-movements of a stonemason whilst at work; but, like all other machines designed on this principle, was not permanently successful.

In the early patterns of Mr. Holmes' stone-dressing machine, the face of the stone to be dressed is placed in a vertical position, and acted on by cutters, which are mounted in a cutter-stock, carried by travelling arms. The cutting-tools are arranged in pairs, and by means of a lever a rocking motion is given to the cutter-stock, causing the right and left-hand tools to act alternately on the face of the stone. The lever is actuated by an eccentric of adjustable throw on a vertical shaft, carried by the outer ends of the travelling arms, and driven by a bevel-wheel which traverses on the main shaft. The feed-motion is given to the travelling arms by two longitudinal screws driven by a pawl and ratchet-wheel. The pawl is reversed at each end of the traversing motion, so as to reverse the direction of the feed. The stone is held in position by clamps and wedges, and by inclining the bed-plate or table on which it was mounted, surfaces could be dressed at oblique angles to one another.

Mr. G. Hunter's first machine for sawing and dressing stone was patented in the year 1855, but we have already noticed it at length.

Schwartzkopff and Phillippson, in the year 1859, patented the adaption of the principle of the steam-hammer for dressing stone, adjusting the blows of the hammer to the nature of the material under operation. The steam cylinder was mounted by the inventors on a sliding bracket or frame which traversed up and down the face of the main framing by raising or lowering the sliding frame; the piston-rod which carried the cutting tool was either lifted from, or brought closer to, the stone being operated on, at the option of the workman. The steam for working the piston was admitted to or shut off by a rotating valve, actuated by two cams (carried by a cross-head attached to the piston-rod), which acted on a projection on the spindle of the valve, and caused it to rotate. In order that the steam-pipe which admits steam to the valve-box of the cylinder may be moved up or down when the cylinder is moved, its end passes through a stuffing-box in the top of a chamber in the upper part of the main frame, into which chamber the steam is first admitted. In dressing stones the whole apparatus was mounted on a slide with horizontal and vertical movements similar to those found in a planing machine for working iron.

The stone was mounted on a truck running on rails, and brought beneath the cutting-tool by means of a worm and rack worked by a hand-wheel. After each blow of the piston-rod containing the tool, on the stone, it was forced back to its original position by means of springs. We believe this machine has never been practically used in this country, but we think that the plan is sufficiently original, and of enough promise to merit further experiment in this direction. For dressing granite or other very hard stone, if the cutting-tools could be made to stand, this machine should be as successful as any other that has been tried for this purpose; but as most machines for dressing granite have hitherto been partially or wholly failures, this is perhaps not very high praise, and there is no doubt a really practical and commercially successful machine for dressing granite yet remains to be invented.

In the year 1860, Mr. C. Schiele, of Bebbington,

patented a machine adapted for cutting or dressing-stone, amongst other operations. The general arrangement of the machine may be described as follows:—On a revolving shaft was fixed a large boss or centre-plate, on which were linked, but free to swing, a number of arms, and when employed for dressing stone a cutting hammer was mounted at the end of each arm. On the cutter spindle being set in motion, the arms were thrown out to a radial position, from the centrifugal force imparted, and operated on the stone blocks to be dressed, which were fixed on a truck arranged to travel beneath the cutting-tools. The cutting tools employed were fashioned after the manner of those used by stone-masons—whose actions they were intended to imitate — for dressing purposes; each of these tools was fixed by its handle, strengthened by a taper hoop, holes being drilled through the handle for the tool to swing in, round a pin secured to another hoop of wrought iron. A number of these hoops were fixed side by side on a shaft in such a way that when the shaft was put in motion they delivered their blows on the stone in succession; each hoop was provided with an elastic cushion. To prevent injury from the tools flying off—which we should imagine they would be likely to do—guards were placed over them, and small jets of water were used to keep the tools cool and lessen the dust. We believe this machine was unsuccessful—never came into practical use; but as an attempt to imitate the hand-labour of the stonemason by mechanical means, it is worthy of note.

A machine for dressing stone was exhibited by Mr. Jos. E. Holmes, of Mold, at the International Exhibition of Vienna, 1873; its general arrangement was somewhat similar to a planing machine for iron. The table was traversed on rollers, and driven by a rack and pinion;

the stone was held in its place by brackets fastened on a table by adjustable bolts. The cutter-barrel or spindle was made adjustable, and two kinds of cutters were employed. For roughing out, a series of independent tools or punches mounted in two rows were employed, and for finishing, a plain blade or cutter the full length of the block was used. By means of a very ingenious vernier combination or disc arrangement, the cutters could be set to take a greater or less cut, according to the angle at which they were fixed.

The system of stone-dressing pursued in the Cook and Hunter patent machines of the earlier types, was to mount the cutting tools spirally round an axle or cylinder, the shortness and sharpness of the tool-cut being regulated by the diameter of the axle or cylinder, the number of cutters, and the rate of feed. For face-dressing the softer kinds of stone, this system has some advantages; but for working hard stones, we consider another system of revolving circular cutters, arranged with a rolling contact, altogether superior. It can however, be used with advantage in moulding Bath, Portland, Caen, and other soft stones, and under this head we shall further describe it.

Our illustration, Fig. 11, represents a machine of the horizontal-barrel class from the designs of Mr. Holgate. As will be seen from the drawing, two cutter barrels are arranged, the first carrying narrow cutters about $\frac{3}{8}$in. wide for roughing out the face of the stone, and the second barrel carrying cutters about 1$\frac{3}{4}$in. wide for finishing the surface. The cutters are forged from bar steel 1$\frac{1}{2}$in. by $\frac{3}{16}$ and 1in. by 1in. respectively. The travelling table which carries the stone is arranged with a quick return motion. The cutter barrels are arranged to be raised and lowered either by belts or hand as may be preferred.

FIG. 11.—DOUBLE HORIZONTAL BARREL STONE-DRESSING MACHINE.

To face page 62.

A few years since, a machine for dressing stone was patented by Messrs. Brealey and Marsden. The general arrangement of the machine is similar to an ordinary planing machine for iron. On the horizontal cutter-shaft are mounted a number of eccentrics—apparently in a very similar way to that employed by the Earl of Caithness many years before—all having the same throw but arranged spirally round the shaft. Each eccentric carries two rods, which are set at an angle of 45°; to each of the rods small brackets are bolted for carrying the cutters, which are made of thin steel plates of different forms, according to the work to be performed. Slots are made in the cutters for adjusting them, but they are not allowed to project far for fear of springing. Two sets of cutters are employed, one set serrated for roughing out, and the other set plain. Either of these sets of cutters can be brought into operation by means of a rocking quadrant frame worked by a hand-wheel, and the cutters set to meet the stone at any desired angle. The cutters operate by a rapid succession of blows across the slab.

The general arrangement of this machine is massive and well designed, but its weak point is, without doubt, the large number of cutters employed.

The most efficient machines for dressing plain surfaces on hard stones, with which we are acquainted, are those made under the patents of Messrs. Brunton and Trier. These are based on principles differing essentially from other machines in use, and from their novelty are deserving of an extended notice. The pith of the invention, as described by the patentees, consists in giving to circular cutters a determinate rotation on their own axes at the same time that they are carried round in a circle, their cutting-edges describing a circular path. The aim of the inventors has been to so adjust the rates of cutter rotation and movement round the circle, relatively the one to the other, that the cutting edge shall exactly roll in the circular path or track. The patentees claim that with a nicely-adjusted rolling action there is very little attrition, and that this is chiefly due to the forward movement of the stone; at the same time little heat is produced and the cutter-edge wears away very slowly. This latter premise must, however, of course, depend on the nature of the stone, as should they have to pass through pyrites, or what quarrymen call "flock," the wear and tear of the cutters must of necessity be largely increased. The velocity at which the cutters are speeded for ordinary purposes is 2,000ft. per minute; but this may be increased according to the nature of the stone being operated on, as the inventors' experience has shown that the greater the velocity of the cutters the better they act.

In an interesting paper on their stone-dressing machinery,* Messrs. Brunton and Trier claim that their machine, as contrasted with others in use, takes hold of a new principle of action—the action, namely, of circular rotating cutter, operating by rolling to chip off from the stone the inequalities of its surface. This, they say, constitutes the elementary principle, and may be stated as a rolling pressure brought to bear at the base of a certain projecting portion of stone, with the intent to force it off. The great power of such a pressure to effect the desired object, is due to the fact that its incidence at any given moment (or what may be called the tread of the cutter) extends over a very small space, and that upon this small space the whole force in exercise is concentrated.

We have in stone a material composed for the most

* Read before the Institute of Mechanical Engineers, January, 1881.

part of particles hard enough to cut and wear away the hardest steel, but held together by a cohesion relatively far feebler than that which holds together the molecules of steel or chilled cast iron. Hence it will be evident that in attacking such a substance by a metal tool, it is of the first importance that attrition be avoided. If this enters in any considerable measure into the conditions of the contest, the metal will be worsted; but if it be a question of simple pressure, the stone will inevitably be overcome. The first application of the principle was to the turning of stone, especially granite. The simplicity of this application was due to the circumstance that the constantly-revolving stone presented a continuous surface for attack; and the contact of the edge of the rotating cutter with the surface was therefore unbroken. The cutter, once set in motion by contact with the stone, continued rolling, and, being placed at an angle of about 25° to the axis of the stone, chipped the surface away incessantly in a spiral line, as the slide-rest and tool-holder moved along the bed of the lathe. The concurrent revolutions of the stone and the cutter reduced attrition to a minimum, and considerable speed of surface rotation was attainable. With two cutters, one on each side of the column, an inch and a half or more would be taken off in a single traverse. But when plain surfaces had to be dealt with, many difficulties presented themselves. The contact of the cutters with the stone was necessarily intermittent. To accomplish a useful quantity of work speed was required; but to bring cutters into rapid rotation by a contact with the stone, which was made and broken at every moment, involved much attrition, and consequent wear. Although it may seem a very simple remedy for this difficulty to drive the cutters—in other words, to give them mechanically an independent rotation, such that their edges should roll on the stone—yet this simple remedy was not thought of till a considerable time had been spent in efforts to dress plain surfaces by simple contact.

An essential feature in the machines under notice is the construction of the tool-holder or chuck employed. These are arranged to carry three or more circular saucer-shaped revolving cutters, made of chilled cast-iron or steel. Each cutter is mounted on a spindle, and set at an angle to the plane of its rotation. These spindles carry also small toothed wheels, which gear into a much larger one mounted in the centre of the chuck, and which imparts to the smaller toothed wheels, and, therefore, the cutters, a rotary motion. The chuck itself is carried on a hollow shaft, through the interior of which passes a second shaft or spindle, on to which the before-mentioned large toothed wheel is keyed. Motion can thus be imparted to the chuck and cutter in proportion to their relative diameters, and the speed thus varied according to the nature of the material being operated on. The cutters may be inclined either towards or from the centre of the chuck, thus giving rise to two types of chucks, internal and external, the internal cutting with the inner edge of the cutters, and the external with the outer edge. Each type has its own merits, and we illustrate them herewith (Figs. 12 and 13) They are thus described by the inventors:—

Fig. 12 represents an internal chuck, having the cutters inclined towards the centre. If the stone can be brought within the circle of the track of the cutters, it may be moved in either direction against them. If the stone is too large for this, the bed must be inclined to the plane of the chuck sufficiently to allow the rough undressed surface of the stone to pass, as shown at a. Chucks of

STONE-DRESSING AND PLANING MACHINES. 67

this type are usually of considerable diameter, and are chiefly used for dressing large blocks.

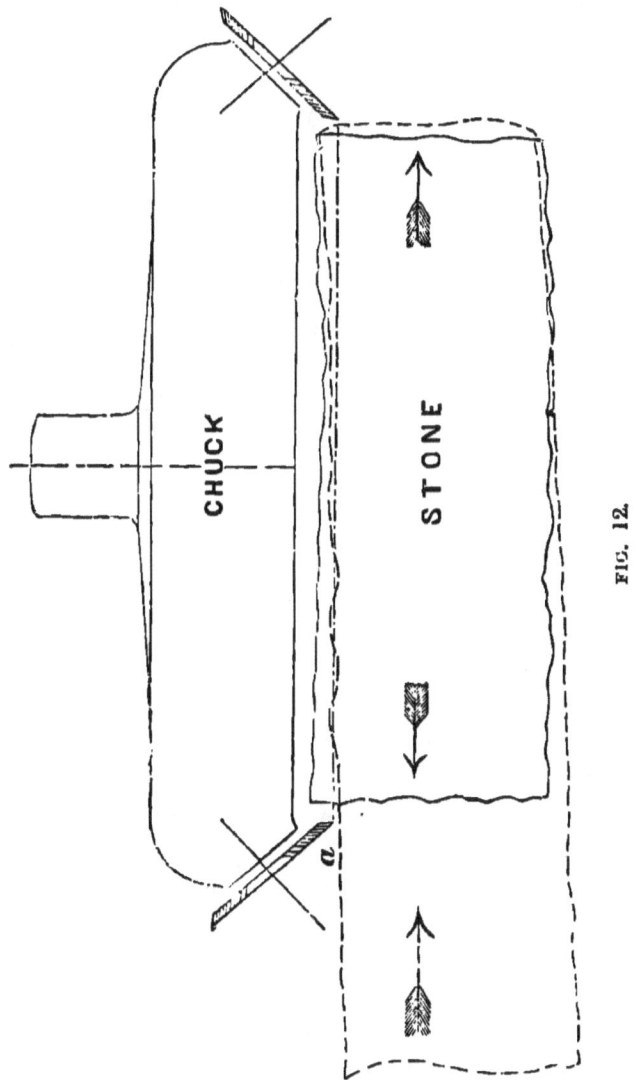

FIG. 12.

Fig. 13 represents an external chuck, that is, having the cutters inclined from the centre. The stone moves from the outside of the chuck against the revolving cutters. An inclination of $\frac{1}{16}$ of an inch in the diameter of the chuck, to the plane of the dressed surface, is given to the axis of the chuck, in order to obtain back clearance. The inclination or tilt given to the chuck is too slight to produce any perceptible hollow in the

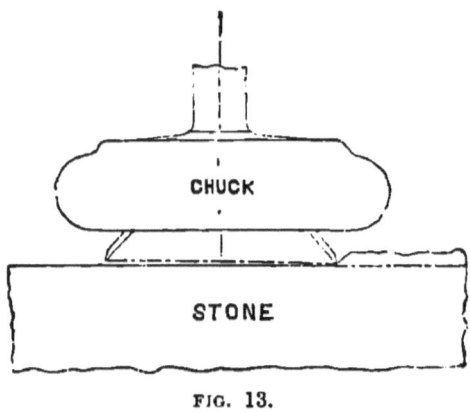

FIG. 13.

surface of the stone, and it is reversible, permitting the stone to travel in either direction. This type of chuck is usually employed for dressing stone of moderate size; but wide surfaces may be dressed in successive breadths by moving it across the face of the stone.

To produce a sharp arris, it is necessary that the line of the centre of the chuck be outside the edge of the stone (see Fig. 14); and to attain the best result, that the cutters roll off the stone, as shown by the arrows, and not on it.

The cutters both in internal and external chucks are set so as to cut in three or more planes or steps (see Fig. 15). When set in three planes, the cutters are distinguished as x, y, or z, the upper or commencing cutter being x, and z the finishing. They follow one after the other in rotation as the chuck revolves, each being on its

STONE-DRESSING AND PLANING MACHINES. 69

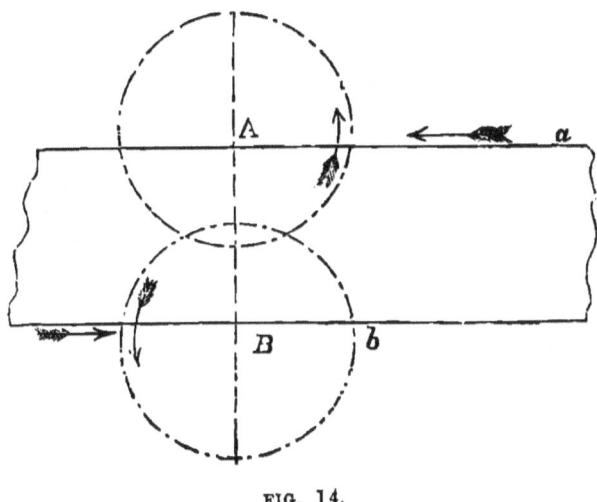

FIG. 14.

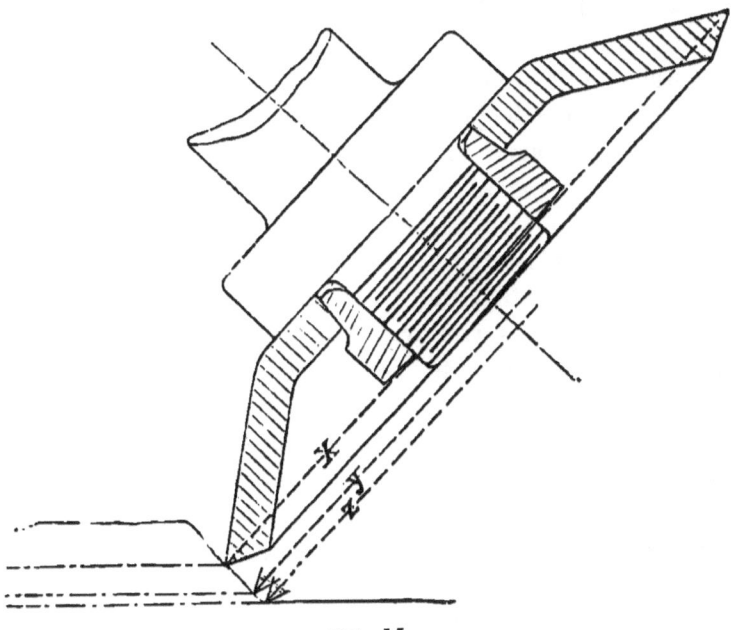

FIG. 15.

own separate spindle. The advantages claimed for this arrangement by its inventors are that a considerable depth of cut may be taken at each passage of the cutters, and thus ordinary quarry-scabbled stone may be dressed at one operation. It is claimed that "plucking" of the stone is avoided, as although the chips forced off by x may break away below its plane, it is unusual in the case of y, and practically impossible with z, owing to the overlying stratum of stone, and any pluck marks that may be produced by x are obliterated by z. It is also claimed for this principle of working that the face pressure of the cutters is greatly diminished, and very sharp arrises are obtainable.

Our illustration, Fig. 16, represents a machine for top-dressing hard stones, such as those usually employed for flags, landings, steps, &c. As will be seen from the engraving, the chuck employed is of the external type, and is arranged with a self-acting vertical and transverse traverse. The stone to be dressed is placed on a travelling table, which passes beneath the cutters, and is actuated by means of a rack and pinion. The main framing of the machine is made of massive construction to overcome the vibration in working. Many difficulties have been surmounted by the inventors in developing this principle of working, and, from samples of stone dressed on their machines that we have recently inspected, the surface leaves little to be desired. As regards the adaptability of these machines for dressing all kinds of stones generally used in building construction there can be but little doubt, but as to dressing the very hardest stone, such as granite, we are of opinion that many difficulties have yet to be overcome, if not an entirely new principle of action invented, before granite dressing by machinery can be made commercially successful.

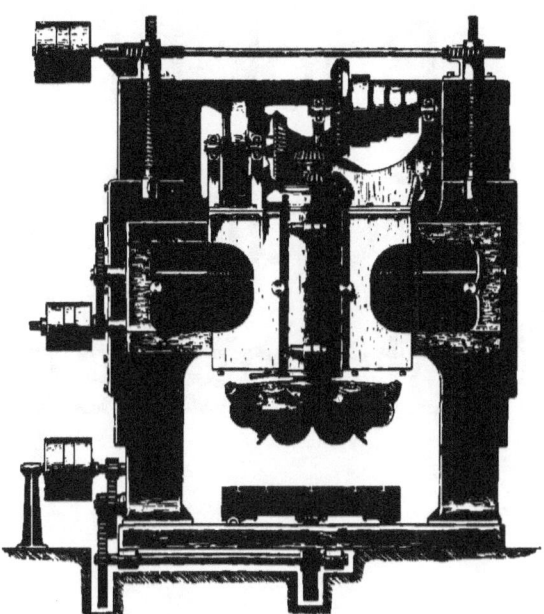

FIG. 16.—REVOLVING CUTTER MACHINE FOR TOP-DRESSING STONE.

To face page 70.

In all classes of stoneworking machinery, where the pressure on the cutters is heavy, or where sand and grit has a chance of getting on to the working surfaces of the bearings, especial means of lubrication must be taken. In the machine last noticed, the chucks, with their spindles and gearing, are lubricated by means of soap and water, introduced through the central shaft, which is made hollow for this purpose, and thence passes to the various working parts.

Fig. 17 represents another type of machine especially adapted for dressing the sides of stones of a moderate degree of hardness, as generally used in building construction, for ashlar mullions, copings, sills, &c. Long stones can be dressed on the ends and jointed by them; and it possesses the advantage of cheapness over the machine already described. The chuck employed is of the internal type, and dresses the whole surface of the stone at one sweep, and the table carrying the stone is speeded to travel from 10in. to 20in. per minute. The chuck revolves in a fixed position, but can be adjusted laterally against the face of the stone.

The inventors construct another machine of the same type as the one last described; but instead of the chuck revolving in a fixed position, it can be raised or lowered automatically, so that stones may be dressed in successive breadths. When it is necessary to dress pillars they are mounted in cradles or fence brackets bolted to the table, and these are arranged to give any desired angle to the stone.

Messrs. Coulter and Harpin patented, in 1872, improvements in machinery for dressing or facing and shaping stone, marble, or granite. The improvements claimed are the use of straight cutting tools, which are so arranged that they turn over after every single traverse of the table carrying the stone, and present a fresh cutting edge; they are thus kept constantly sharpened, and

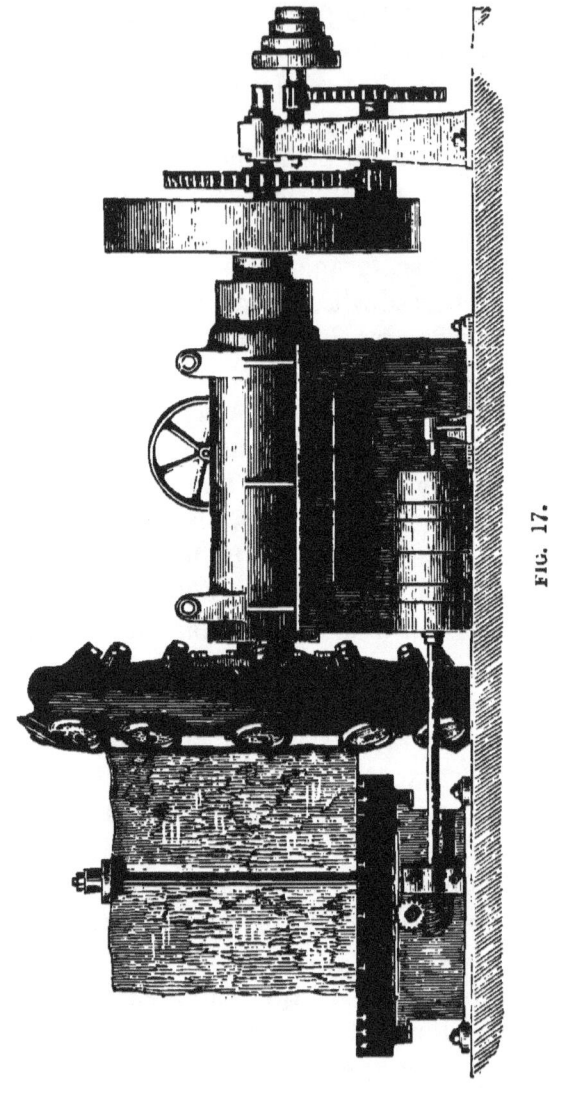

FIG. 17.

operate on the stone during both the back and forward journeys of the table.

FIG. 18.—HORIZONTAL CUTTER STONE-DRESSING AND MOULDING MACHINE.

To face page 72.

As will be seen from the illustration, Fig. 18, an ordinary planing machine for working iron is the basis on which their improvements are founded. The cross-slide bed is mounted on an axis in brackets, which can slide up and down the upright standards. The cross-slide bed is slotted to receive a tool-box, thus bringing the tools through the centre of the bed, and consequently the pressure on the tools has no tendency to lift or move the cross-slide during the cut. The tool-box is arranged with a transverse movement by means of a screw in the ordinary way. The turning over of the cross-slide bed for reversing the position of the cutting tools is made self-acting in conjunction with, and actuated by, the reversing motion of the travelling table carrying the stone. In order to check the fall of the cross-slide bed when turning over, an air-cylinder and piston are used, the cylinder being charged with air, which is compressed by the fall of the cross-slide bed, the air then passes through perforations in the piston, thereby allowing the full weight of the cross-slide bed and tool-box with tools to act as a resistance against the pressure on the tools. The travelling table carrying the stone runs on angular-shaped rails, so shaped to avoid an accumulation of sand or grit, as would be found were the ordinary V-shaped slides used. The general arrangement of the machine is well thought out, and taken altogether it must be considered to be a good example of its type. For working curbs and channels, mouldings and general mason's work, a rocking table can be employed with advantage, as the stone, when once properly fixed, can readily be set to any desired angle, or turned any side up, as may be required. Fig. 19 represents a tool-holder and tool.

In some stone-dressing machines of recent construction, in place of the ordinary fixed tool-box and tools employed in a planing machine for working iron, one or more cutter-barrels are arranged horizontally. The cutters are usually forged from plain bars of steel, and if two barrels are used it will be found best to arrange the first barrel with narrow cutters for roughing out, and the second barrel with broader cutters for finishing. The cutters should not project from the barrel more than three or four inches, or the leverage and consequent attrition on the cutters will be largely increased; and it will be found advantageous to arrange them spirally on the cutter-barrels so that the blows on the stone and consequent jar and vibration will be distributed, and not be one sudden concussion across the whole length of the stone, as would be the case were they arranged in straight lines.

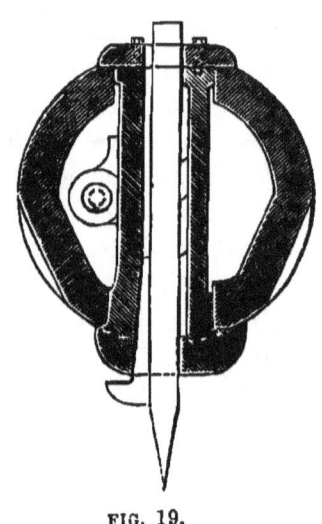

FIG. 19.

But any machine of this type, employing a number of cutters, as we have before remarked, labours under the serious disadvantage of inequality of wear in the tools. The cutter-barrels should in all cases be raised by power, as much time is lost when manual labour is employed. As the pressure and strain on the bearings carrying the cutter-barrels is very considerable, they should be made of increased length, say about three diameters, and preferably made of phosphor bronze. The travelling table should be fitted with a quick return motion, should run on rollers, as the ordinary V-slides would be rapidly ground away by the grit and dust made in working.

Plain disc surfacing machines for stone have been used with some little success, especially for tolerably soft stone. They consist briefly of cast-iron discs arranged to revolve vertically or horizontally. Around the periphery of the discs is arranged a series of cast-steel cutters which act on the stone. The stone is mounted on a table, and traversed either past or beneath the cutters. Rectilinear reciprocating machines for dressing stones have also been tried; in these the cutting tools are mounted in a traversing slide, and are arranged to operate on the stone at every traverse of the slide. For scabbling large blocks from the rough, this latter arrangement may be used with advantage in the quarry; but for finishing purposes, it is hardly good enough. The tools are generally made with double noses, so that they may cut both ways of the traverse. For dressing ashlar, or for bedding stones in lieu of a rubbing bed, plain flat steel cutters fixed in movable cutter-blocks, mounted on horizontal or vertical revolving spindles, may be used with advantage. The travelling tables of planing or moulding machines should in all cases be arranged with variable rates of speed, to suit different classes of stone.

The dressing of granite by mechanical means is a problem that has engaged the attention of engineers for many years, and is still, practically speaking, unfulfilled. The invention of rotating cutters for turning and dressing granite is generally, and, I believe, rightly attributed to a Mr. Newton, who spent large sums of money in constructing machines for granite dressing, which were, however, not practically successful. Mr. Wm. Johnson, of the Hayton Granite Co., also took out a patent about 40 years ago, with the same object in view, and a number of others have since been taken; but we cannot learn of any one of them being a success. The nearest approaches, however, are those of Messrs. Brunton and Trier. Fig 20 represents Trier's Patent Cutter-chain Vertical Stone-dressing Machine.

The machine illustrated is a Side Dresser, to which the following description chiefly applies: The circular cutters, A A A, are fixed to spindles, which are free to revolve in adjustable bushes carried on the boxlike slides, B B B. These slides are hinged together, and form the links of an endless chain, which passes round sprocket wheels, borne on the standard, D.

Guides are provided on each side of the standard, through which the slides or links of the chain pass, so that the cutters are firmly guided in straight lines, both ascending and descending, and as they come into contact with the advancing stones, and split off the irregularities by their rolling wedge action, they produce good surfaces, with straight toolmarks at right angles to the length of the stone. By raising one end of the stone, diagonal marking to any desired angle will result.

The Side Dressers are provided with a bed, E, on one side or on both sides of the standard, according to the desire to dress one stone or two stones simultaneously. The table, F, to which the stone is fixed, with the side to be dressed overhanging the edge, has wormrack, slides and wheels. The table is rolled on rails up to the bed, on to which it is drawn by a wormshaft, and by which it is pushed off on to rails at the other end. By means of cross trolleys and return rails, two or more tables can in this way be circulated, and the loss of time involved in changing stones is practically done away with.

The cutter-chain, with its considerable number of cutters moving continuously in one direction, permits the valuable principle of stepcutting with several cutters to be carried out without sacrificing the speed of dressing.

FIG. 20.—TRIER'S PATENT VERTICAL STONE-DRESSING MACHINE.

To face page 76.

CHAPTER VIII.

STONE-MOULDING MACHINES.

The art of moulding stone by manual labour is of very ancient origin, but of the implements employed, and their nature and composition, comparatively little is known. Wilkinson is of opinion that the action of cutting and grinding the faces and making ornaments in relief or intaglio was performed with emery, probably imbedded in some soft metal. The early form of cutters used for stone or marble moulding by machinery, consisted of solid cylinders of cast iron, turned to the counterpart form of the required mouldings. These were mounted on horizontal or vertical spindles, as required for flat, circular, or edge moulding, the table being traversed beneath or past the cutters, as in the machines at present in use. Amongst the early machines must be mentioned the old ridge roll which has been used for many years for rounding the edges of slate slabs.

In the year 1833, as we have elsewhere mentioned, Mr. G. W. Wilde took out a patent, and in this was included some improvements in moulding stone by means of sand and water, acted on by a metallic cylinder on the periphery of which was formed the converse of the intended moulding.

The old form of machines were termed "moulding

beds," and several of these were constructed by Mr. James Tulloch. They were constructed somewhat after the fashion of the old ripping bed we have already described; but in place of the disc of sheet iron he mounted a circular cutter block, on the periphery of which was turned the counterpart of the moulding to be produced. The strip of marble to be moulded was fixed with plaster of Paris to a board or table, which was capable of being moved slowly forward by a rope and counter-balance weight. The "faithful historian," in describing this machine, says: "The cutter being set into rapid rotation the marble is brought up to it, and is cut away by the action of the revolving iron. The marble advances onwards as fast as it is cut, and then presents a series of parallel mouldings on its surface, the counterpart of those in the cutter." With all due deference to the "faithful historian," we are afraid that if no other cutting or abrading material was employed, the work turned out on this machine must have been, to say the least of it, rough, or else the marble must have been very soft and the cutters very hard. There is little doubt, however, that a constant supply of water and very fine-grained sand was introduced between the periphery of the cutter and the marble. The writer has recently tried several experiments in moulding marble, with, so far, very encouraging results. In lieu of iron cutters and sand, he has used a certain quality of emery disc or wheel with the profile of the desired moulding formed in its periphery; this he found to cut with considerable rapidity, and with certain modifications in the constituents of the disc, and in holding and feeding the marble, it should be made practically and commercially successful.

About the year 1853 Messrs. Knowles and Bellhouse, of Manchester, patented a series of machines for cutting and shaping marble and stone. In one of these they used a rotary cutter or rubber combined with a template guide for the direction of the shaping surface in the accurate contour line of the intended shape of the article to be produced from the rough block or slab, as it lay upon a traversing table. When symmetrical or circular mouldings were required, the inventors attached a set of shaping or moulding blocks to a revolving disc; this was mounted on a vertical spindle and brought on to the work by means of a lever. In another machine for circular moulding, the necessary working pressure was obtained by means of weights or springs, whilst the table carrying the stone was arranged with rotary and traverse motions, so that, if necessary, it could be made to revolve as well as the cutters. We have drawings of these machines before us, and they appear thoroughly practical, and reflect credit on their designers.

Moulding machines for working stone may be divided into two classes: (1) those in which the cutter spindles or barrels work in a horizontal position. (2) Those in which they work in a vertical position. As we are inclined to favour the latter type, we will discuss these first. In the earliest machines, as we have already mentioned, the converse of the moulding required was turned in cast iron and afterwards chilled. Flat cutters of steel mounted in a vertical revolving spindle followed these. And also a kind of hollow saucer cutter, made of chilled iron, which was attached to the foot of a vertical spindle. This was guided in its action by a pointer attached to the slide which carried the spindle, and by a grooved pattern attached to a portion of the table, which was made to rotate.

In one type of vertical moulding machine, the stone is mounted on a travelling table, and traversed by means of a screw-feed past a vertical cutter-barrel, in which are fixed several sets of roughing tools, usually six—these roughing-down tools are plain rolled bars of steel, of a section similar to Fig. 21. These can be advanced or retired in their tool-box to suit any profile of moulding required. The cutting edge of one set projects a little over the previous set, so that each time the stone is traversed six cuts are made, the result being great expedition in roughing the stone down to the approximate shape of the moulding required. The roughing tools are kept in position by suitable packing pieces, and only one set is required for any section of moulding that may be desired. As soon as the stone is roughed down by these tools to the approximate shape of the moulding, it is scraped to its exact outline, and finished by a scraping cutter, which is fixed in the tool-holder or cutter-barrel on the right-hand side, or behind the roughing tools. This finishing cutter is brought into action, and the amount of its cut regulated by a hand-lever, under the immediate control of the workman. If the roughing cutters are carefully adjusted, little work need be done by this tool, and a finish and sharpness of outline is produced quite unattainable by hand.

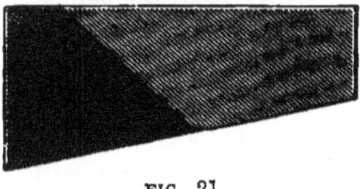

FIG. 21.

The table which carries the stone can either be used horizontally, or placed at an angle which will be found convenient for undercut and other mouldings. The table is traversed by a screw-feed worked by two sets of pulleys, one for moving the table forward whilst cutting, and the other for running it back at a higher speed, preparatory to taking a fresh cut.

Amongst the most advanced machines for moulding stone must be included Hunter's Patent Duplex machine, which we illustrate herewith (Fig. 22). In this machine the stone is subjected to the action of both revolving and scraping tools. The improvements consist in mounting two vertical spindles which carry revolving cutters, one

FIG. 22.

on each side of a reciprocating or travelling table, upon which the stone to be worked is fixed. The cutter spindles are driven by screw or worm gearing; the two worms being inside right and left handed, the thrust of each partly counteracts that of the other. The table is driven by means of a worm working into a rack fixed on the underside of the table; this worm is made slightly taper on a shaft somewhat inclined, so that the outer end of the shaft to which is applied reversing bevel gear is sufficiently low to permit the table to travel over it.

The mouldings are roughed out by the small trumpet-shaped steel cutters similar to those used for circular sawing, and shewn in Fig. 4; these are mounted horizontally

G

on flat malleable iron tool holders, similar to Fig. 24; these are arranged on the vertical spindles in rotation, the sizes of the tool holders being graduated according to the depth of the various members of the moulding.

For scraping and finishing the stone, flat profile steel cutters are used; these are fixed in tool holders, mounted on vertical columns with elevating screws; on

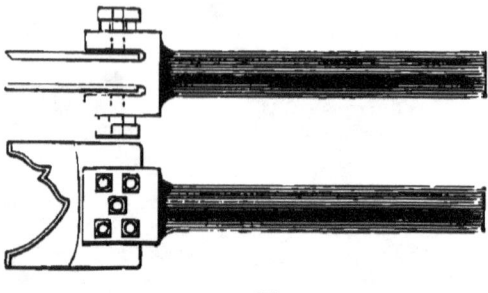

FIG. 23.

the upper part of these columns are fixed cross tubes, in which the tool holders are fixed. These are adjustable horizontally across the travelling table by means of screws, so as to bring the tool to act either on the side or on the upper surface of the stone as required.

The one tool scrapes or planes when the table travels the one way, and the other when it travels the reverse way, so that no time is lost in running the table back for a fresh cut, as in other machines of this class. The top of each tool holder is fitted with a swivel to hold an adjustable cross bar for planing an inclined face.

Mr. Hunter adopts a simple plan for holding the cutting tools; this is illustrated by Figs. 23 and 24. It consists of a casting having holes for the reception of the tools, and transverse steel keys fitted on inclined beds. The tool being inserted into the hole is fixed securely therein by drawing up the key by means of a nut. The key is formed with a V-edge, which slightly impresses itself into the tool, and prevents it from moving. Two or more tools can be fixed in one holder, and each tool is set so as to overlap that in the next holder. These tool-

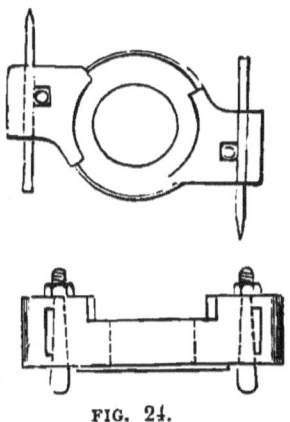

FIG. 24.

holders or blocks are for some kinds of stone mounted in trunnions, thus presenting a fresh cutting edge to the stone when the spindle is reversed. Figs. 25 and 26 show the method adopted by the inventor for mounting

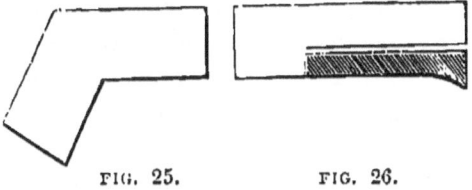

FIG. 25.   FIG. 26.

the scraping tools, which are adjustable horizontally through the cross tubes before alluded to. Fig. 4 represents one form of cutting tool employed by Mr. Hunter. The general arrangement of this machine is compact and good, and the combination of revolving and scraping cutters is not, we believe, found in any other machine of this type. For moulding the softer kinds of building stones this machine should be especially serviceable. For moulding and surfacing semicircular windows and such like curved work, a revolving table is generally employed, the cutters being usually made adjustable in a slide, but stationary, the table carrying the stone being arranged to revolve against them. For working the softer stones this machine can be made to effect a large saving over hand labour.

In moulding or planing stone, it is of the utmost importance that the blocks are very firmly fixed. There are several ways of doing this. If a number of pieces of the same size have to be worked, a good plan is to make a strong skeleton frame of hard wood, and fix it securely to the travelling table. Into this frame the stone can be lowered and wedged up. This dispenses with the ordinary method of cramping, and is more expeditious. To further support the tools and stone whilst being moulded some engineers form the lower part of the box carrying the cutting tools as near as possible to the profile of the desired moulding.

For holding shafts or pillars in position whilst being moulded or planed, a centre head or head stocks of modified form are usually employed. These can be moved about on the table, and made adjustable to blocks of varying lengths. The stone can also be fixed at any desired angle.

When working mouldings, a template of the profile of the required moulding should, in all cases, be made in sheet iron and marked on the stone. The cutting tools can then be more readily adjusted to suit.

In dressing and moulding machines where a number of cutters are employed, or where they are difficult to adjust, it will be found well to sharpen them in their places. This can usually be done without much trouble, by means

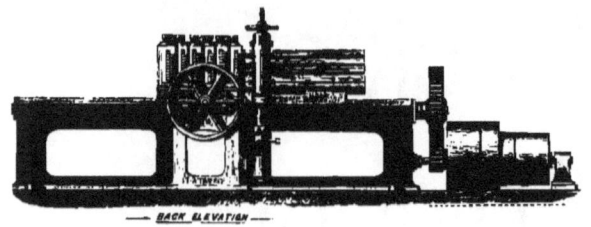

FIG. 27.   FIG. 28.

FIG. 29.—VERTICAL BARREL STONE-MOULDING AND PLANING-MACHINE.   *To face page 55.*

of a revolving emery wheel, mounted in centres, and driven by an elastic band.

In making bevelled mouldings the stone should, if possible, be sawn to an angle before being put on the machine, as the cutters having thus less stone to remove are subjected to less strain, and consequent wear; at the same time the stone is less liable to pluck in the working, or break at the arrises. For "checking" deep mouldings, a circular saw may be used with advantage, this is fixed on the top of the other cutter holder, and cuts into the stone before the cutters following it come into play. In moulding stone, all complex forms of cutters should be avoided; and many members of the same moulding should not be formed on one cutter—except for scraping and finishing purposes—or considerable trouble will be found in keeping the iron to the desired outline : at the same time, it is much more liable to twist or crack whilst being hardened. The temper of the cutters should be adjusted according to the hardness or nature of the stone being worked. We can recommend the following plan for hardening cast-steel tools, the edges of which—provided the steel be good—will be found to stand well : Take four parts of powdered yellow resin and two parts of train oil, mix them carefully, and add one part of heated tallow. The object to be hardened is dipped into this mixture red-hot, and is allowed to remain in it until it is quite cold. Without having previously cleaned it, the steel is again put into the fire, and is then cooled in boiled water in the ordinary manner. We purpose giving elsewhere some further directions as to tempering the cutting tools; the subject being one of the utmost importance in securing the efficient working of the various machines.

Our illustrations (Figs. 27, 28, and 29) represent a stone moulding and planing machine from the designs of

Messrs. Western & Co. The cutter barrel is a fixture and arranged vertically, and plain bars of steel of taper section first of all rough out the profile of the desired moulding, which is afterwards scraped smooth by a steel cutter formed to the exact outline required. In working soft stone, such as Portland, a considerable amount of stone can be removed at each cut. The stone to be worked is fixed on a table, which is traversed by means of a screw past the cutters till the moulding is shaped to the desired outline. It will be found well adapted for dressing steps, string courses, sills, and other builders' stonework. The table carrying the stone is fitted with a false top, hinged in front, and adjusted by screws, so that mouldings can be worked whose outline would otherwise be undercut from the face of the tools. The tools are readily adjustable, being divided by packing pieces, and held firmly in position by set screws.

Those stone-moulding machines in which the cutter barrel is arranged to work in a horizontal position, have hitherto been modifications of either planing machines for working iron, or circular stone-sawing machines. In one type of machines the cutter-barrel is made to revolve, the cutters being fixed in its periphery, whilst in another the cutters are fixed in a stationary tool-box after the same manner as tools used for planing iron. For roughing out mouldings a number of small saws, varying in diameter according to the profile of the moulding, are occasionally employed, instead of specially shaped tools; they are, however, too expensive for general use.

The travelling table which carries the stone can be arranged with a reversing motion, and by the employment of double-nosed tools they may be made to act on the stone during both passages of the table. Two tool boxes may be employed—one on either side of the table—the one being used to finish a moulding, whilst the other is roughing out. For working architraves, strings, cornices, &c., in soft stone, the fixed tool boxes used in this and the vertical type of machines should be useful; but for facing or heavy moulding purposes, especially with stone of any degree of hardness, we do not approve of the system, as the dead contact of the tools and the stone is too prolonged. The framing of this type of machine should be of massive section to overcome the vibration in working. In working soft stone by this method, the table should travel at about 4in. per minute for roughing out, and about 10ft. per minute for finishing with the scraper; these speeds, however, should be modified according to the nature of the stone being worked. By fixing—in addition to the moulding cutters—two circular saws on the cutter-barrel, the sides of a block of stone may be squared up at the same time as the moulding is being cut.

For edge moulding and trueing up slabs, &c., vertical spindle machines somewhat similar to those used for moulding and shaping wood have been used with some success. These machines consist of one or more vertical spindles which project above the surface of a cast-iron table; feeding tables which carry the stone or marble are arranged on each side of the machine, and bring the stone under the action of the cutters. The cutters are usually made of steel, and of a shape converse to the moulding required, but vary according to the nature of the work, some being arranged with perforations for supplying emery or other abrading material. The stone to be shaped is very securely clamped on the table, and slowly traversed past the revolving cutter, usually by hand. The travelling table is generally arranged with two motions, one tangential to the circle described by the revolving cutter, and the other at right angles to this. The bottoms of the vertical spindles run in footstep bearings; these should in all cases be made adjustable for wear, and especial means taken for their lubrication. The cutter spindle should be made of steel, and the main framing of the machine of substantial section, to overcome as far as possible the vibration of working.

Our illustration (Fig. 30) represents a machine for moulding and planing stone by Messrs. Rotheroe, Sherwin, & Co. The machine, as will be seen from the sketch, is an adaptation of an ordinary planing machine for iron; the travelling table is fitted with reversing motion, and two double boxes are fitted which can carry, if required, two tools, so that the stone may be, if necessary, operated on during both traverses of the table. The stone is roughed out with a double-nosed tool and finished with a steel cutter shaped to the exact profile of the required moulding.

In concluding our remarks on stone-moulding machines we may say we are decidedly in favour of using a combination of revolving cutters for roughing out, and stationary scraping cutters for finishing, in preference to two stationary sets of cutters; the dead contact of the roughing cutters with the stone being less, consequently the wear and tear and power required to drive is less in proportion and there is less liability to " pluck."

FIG. 30.—HORIZONTAL FIXED CUTTER MOULDING AND PLANING MACHINE.

To face page 58.

CHAPTER IX.

RUBBING OR SURFACING BEDS.

RUBBING beds usually consist of large discs of cast iron mounted on the top of a vertical spindle, to which a rotary motion is given by suitable toothed gearing.

The rough stone to be faced is placed on this disc and furnished with a supply of sand and water. The disc being set in motion, a perfectly true and smooth face is formed on the stone, which is kept flat on the disc usually by its own weight, and prevented revolving by means of a fence or fences placed across the table, and one or more pieces of stone may be faced at the same time.

Mr. Tulloch's grinding bed differed considerably from those now commonly in use: in his machine the slab to be ground was placed horizontally on a moving bed, and the grinding produced by sand and water, by means of a large flat plate of iron resting upon the surface of the slab. The two surfaces were traversed over each other with a compound motion, partly excentric and partly rectilinear, so as to continually change their relative positions.

The machine consisted of a frame about 9ft. long, 6ft. wide, and 8ft. high. About 2ft. from the ground is mounted a platform that is very slowly reciprocated horizontally for a distance of from 1 to 2 feet, according to the size of the slab, by means of a rack and pinion placed beneath, and worked alternately in both directions. Above the platform are fixed, vertically, two revolving shafts, having, at their upper extremities, horizontal toothed wheels of equal diameter, which are driven by means of a central toothed wheel hinged on the driving shaft. The two vertical shafts are thus made to revolve at equal velocity, or turn for turn; and to their lower ends are attached two equal cranks, placed parallel to each other, the extremities of which, therefore, describe equal circles in the same direction. To these cranks the iron grinding plate or runner is connected by pivots, fitting two sockets placed upon the central line of the plate. The cranks are made with radial grooves, so that the pivots can be fixed by wedges at any distance from the centre of the crank. When the machine is put in motion, the grinding plate is thus swung round bodily in a horizontal circle of the same diameter as the throw of the cranks, which is usually about 12in., and, consequently, every portion of the surface of the grinding plate would describe a circle upon the surface of the slab if the latter were stationary. But by the slow rectilinear movement of the platform, the slab is continually shifted beneath the plate, so as to place the circles, or rather the cycloids, in a different position; and it is only after many revolutions of the cranks that the same points of the surfaces of the grinding plate and slab are a second time brought in contact. The grinding plate is raised for the admission of the slab by means of four chains suspended from a double lever, and attached to the arms of a cross secured to the centre of the upper surface of the plate, which is thus lifted almost like a scalepan. For slabs that are much thicker or thinner than usual, the principal adjustment is obtained by the removal or addition of separate beds or loose boards, laid upon the platform to support the slab at the proper height. Slabs that are too large to be ground over the whole surface at one operation are shifted once or twice during the grinding. The weight of the horizontal plate supplies the pressure required for grinding; and the pressure can be regulated, if necessary, by a counterpoise weight attached to the double lever. The sand and water required for the grinding is thrown upon the grinding plate, which is pierced with a number of holes, and is surrounded by a ledge, so as to form a kind of shallow tray. The sand and water find their way beneath the plates through these holes, and gradually work their way out at the edge.

About the year 1859 Messrs. Coulter and Harpin, of Huddersfield, erected several machines for rubbing building stones. The machine consisted of a circular iron table on which the stone was laid; this table was made to revolve after the manner of the circular rubbing beds now in use, and on the top of it was placed an iron frame, arranged to rise and fall according to the thickness of the stone on the table. Stones were fixed loosely in this frame, and it was made to traverse backwards and forwards over the face of the stone fixed on the moving table. Water and sharp sand were used, and the two faces of the stones worked at the same time. It is reported that at the trial of this machine the revolving table was packed with rough building stones to the extent of about 80 square feet, and the upper frame was filled with paving stones, as they came out of the quarry; the machine was set in motion and ran for twenty-five minutes, when it was found that both sets of stones were smooth and level, more especially the top stones, which were stated to have been as smooth as ice. The amount of stone rubbed at a trial in Glasgow in less than half an hour was 150 super. feet, which is about equivalent to the work of eleven men for one day.

As the stone approaches completion on the rubbing bed, the sand is changed gradually for finer kinds, and if a very smooth surface is required, the finish is usually given with the finest silver sand. If the stone is carefully handled, the work turned out on a rubbing bed will be found far superior to hand labour, and it can be produced at a much less cost. Another advantage accruing from its use is that the mason can make a much neater and closer joint, and in less time than with hand-prepared stone. For facing ashlar, rubbing beds will be found extremely useful, as the rubbing does not bruise the stone as is often the case with mason's hand work. The pores of the stone are also closed by the rubbing, and its general appearance is superior to hand work. The cost of rubbing stone may be set down at from 1d. to 3d. per foot, according to its nature. Heads, sills, steps, &c., can also be readily faced on a rubbing table, and a number of small pieces may be wedged together.

For rubbing or facing long pieces of stone, some masons prefer to mount the stone on a traversing table, and rub its top surface with a disc, instead of placing the stone face downwards on an ordinary rubbing-table, as they maintain they can obtain more uniform work. In facing long pieces of stone there may be some truth in this, as it will be readily seen that that part of a large rubbing disc near the periphery will travel much faster, and, consequently, cut quicker than that near the centre; thus more stone is removed from the end of the block near the periphery than that near the centre. A skilful mason can, however, remedy this in a great degree by regulating the supply of sand and water. It is important that the table be capable of adjustment both horizontally and vertically, and in the larger sizes, in addition to the ordinary footstep bearing supporting the vertical shaft, a further support should be given to it rather more than half-way up. This can readily be done by means of a rest, and one or more wrought-iron girders.

The footstep bearing should be made of steel or phosphor bronze, carefully fitted, and especial means taken for its lubrication, or it may give some trouble in working owing to the excessive downward pressure. This bearing should also be made adjustable for wear; if it is not, and the vertical spindle be allowed to sink, the bevel toothed driving wheel and pinion will become too deeply in gear, and breakage will be the result. The revolving bed should in all cases be surrounded by a trough for carrying away the waste sand and water. A constant supply of different grades of sand should be at hand, so that the workman may use that best suited to the work in hand. Adequate means should also be taken for lifting the stone on and off the bed, so that no time is lost in this way.

As horizontal disc rubbing beds are of considerable weight, being made in some cases of as large a diameter as 14ft. or 15ft., to secure their satisfactory working, it is of the highest importance that they are truly fixed on very substantial foundations. These foundations can be of stone, brick, or concrete, preferably the first. Immediately below the vertical spindle a chamber should be formed for the driving gear to run in, but sufficiently large to admit a workman for adjustment and repairs. The foundation bolts holding the footstep pedestal block and pinion shaft should pass entirely through the foundation, as in working a considerable amount of strain is put on these.

In another arrangement for surfacing and polishing

the stones are attached to face plates fitted on tables arranged with a horizontal traverse; the face plates are made adjustable, so that the stones can be advanced to or from each other, the plane surface of the stone being produced by the rubbing of one stone against the other. The traverse speed of the table is arranged according to the nature of the stone being worked, and an output of from 150 to 200 superficial feet per day can be looked for with a stone of moderate hardness.

For facing and polishing stone and marble of awkward section, or when fixed in an awkward position, a flexible shaft (Stow patent) may be used with advantage; this may be used for transmitting rotary motion to a rubbing disc or plate in any direction from the motive power. This shaft is made up of a series of coils of steel wire wound hard upon each other, each alternate layer running in an opposite direction, and the number of wires in different layers varying according to the work for which the shaft is adapted. About one and a half inch at each end of the shaft is brazed solid, and to these solid ends the fittings are attached, the one to receive the revolving tools, the other to receive the power from the pulley inclosing it, which in turn receives its power from a belt. The shaft is inclosed in a case, consisting of a single coil of wire, its internal diameter being a loose fit for the outside of the shaft, covered with leather or other flexible material. This method of construction insures great torsional strength, at the same time entire flexibility at right angles to the axis; it is thus enabled to reach round corners and out-of-the-way places.

For facing and shaping stone, slate, or marble, a circular plate, arranged with a rotary motion by means of bevel gear, is employed. This can be attached to the flexible shaft by means of a clutch. The emery, sand, or other abradant is used solid, in the shape of a disc, and inserted in a turned recessed plate, which is held firmly on the work to be faced by the attendant. When a change of grinding discs is necessary, the plate containing the disc is unscrewed from its spindle and another put in its place. The coarseness or fineness of the discs can thus be varied at will, according to the surface required, or the nature of the material being operated on.

Stone rubbing or facing is due to the action of an abradant or of mutual abrasion, usually with sand, between two surfaces. Various other forms of rubbing or facing tables than those described are in use. In some of these a flat face on one stone is produced by the rubbing or friction of another. A good way to carry out this plan is to mount the grinding stone, and the carriage carrying the stone being ground, on different centres; an excentric motion is thus obtained, and the carriage being moved along longitudinally, all parts of the stone are brought under the action of the grinder; or two stones may be made to face or dress each other, by arranging the upper one in a frame with a rotary motion, and mounting the lower one on a table or carriage with a longitudinal movement. Where heavy friction is not required to produce a smooth surface, the little machine we illustrate, Fig. 32, will be found extremely useful. This is worked by hand, and consists, briefly, of a horizontal rubbing or scouring disc, worked by bevel gearing. The rubbing disc is connected to the vertical spindle by means of a universal joint. The machine is pushed over the surface of the stone by hand; and for scouring and other purposes on a building contract, where large machines are not available, we can recommend it.

Our illustration (Fig. 31) represents a stone-rubbing and facing table from the designs of Mr. E. P. Bastin.

96 STONE-WORKING MACHINERY.

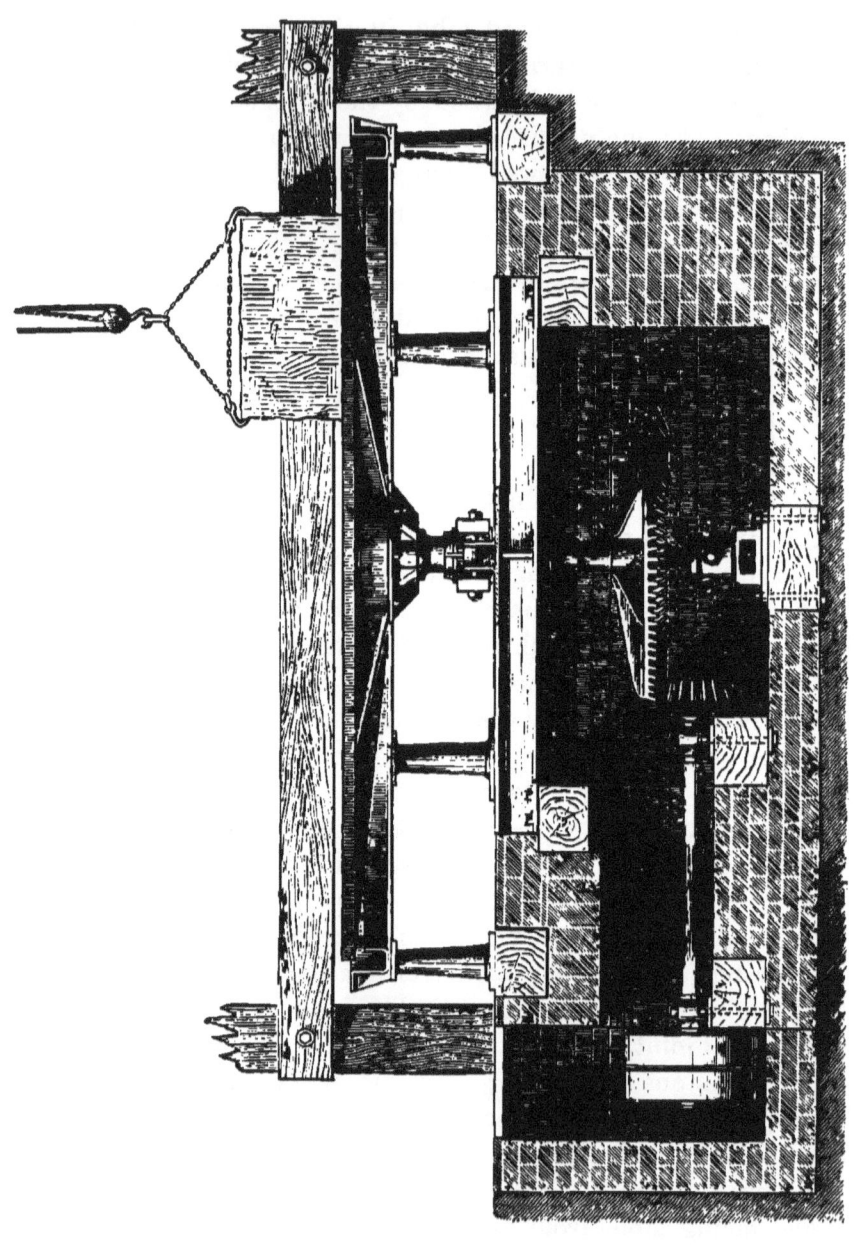

FIG. 31.

Its general arrangement requires little explanation; the centre shaft carrying the table is well supported, and the footstep bearing is made adjustable for wear. A cast-iron

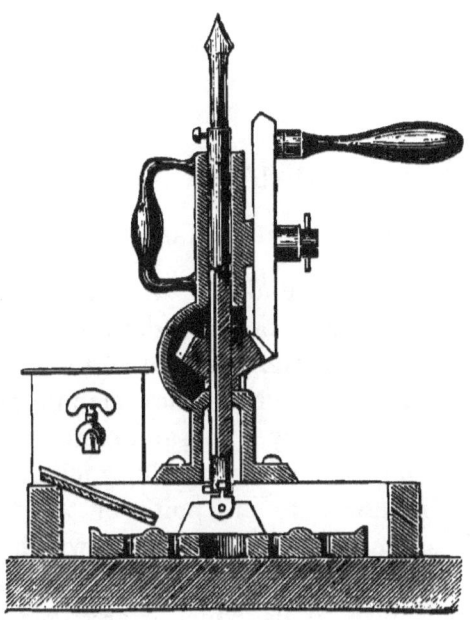

FIG. 32.

trough, supported by cast-iron pillars, for catching the waste sand and water, surrounds the table. The general design of the machine and arrangements for fixing appear compact and good. A set of pulley blocks for lifting the stone on and off the table are arranged overhead.

CHAPTER X.

STONE RECESSING AND QUARRYING MACHINERY.

For recessing and moulding small figures and designs in stone, an overhanging vertical spindle machine, somewhat similar to that used for the recessing and internal shaping of wood, is employed. The vertical spindle carrying the cutting tool is driven by a belt and mounted in a frame arranged with several flexible or elbow joints, so that it may be rapidly moved to any point of the table which carries the stone to be worked, and which is placed beneath it. The stone with an open pattern placed over it is clamped on the table, which is arranged to rise and fall. The cutting tool, which usually consists of a number of diamond points, is mounted on the end of the vertical spindle, and is raised or depressed by means of a lever or handle, as may be required to suit the inequalities of the work, whilst a jet of water is brought constantly to bear on the point where the cutting tool is in operation. This machine is also well adapted for drilling purposes. For moulding large circular slabs four arms, are usually employed; these are of equal length, and are mounted on the overhanging vertical spindle at right angles to one another. At the end of these arms the cutting or scraping tools are fixed; these consist of pieces of wrought iron, bent and shaped to the curvature and outline of the moulding required. By setting the scraping tools nearer or further from the centre, smaller or larger circular mouldings may be produced. The action of the cutters is really an abradant one, the track of the cutters being furnished with a constant supply of sand and water. Either the revolving arms or the vertical spindle must be counterbalanced, so that as the moulding becomes deeper they descend in like manner into the stone or marble.

For cutting out circular holes, wrought-iron cylinders are employed; these are fitted to the bottom of the vertical spindle, and vary in size according to the diameter of the holes required, and the fissure formed by the cutters is supplied with sand and water in the usual way. Diamond points may also be employed for cutting holes with advantage in very hard stone or marble.

For working small irregular or straight mouldings on chimneypieces, edges of shelves, washstands, brackets, marble console tables, and the like, a simple and cheap machine may be made by modifying a wall-drilling machine and mounting a chilled iron cutter block in place of the drill. It will be found advisable to mount the marble to be moulded on a table with a rotary motion, and fixed on a trolley arranged with horizontal and transverse movements. The table can be made to run horizontally in V-slides, the trolley itself tranversly on metals: both of these movements may be regulated by means of chains and counterweights. The marble to be worked must be firmly fixed on the table, and a template of the required moulding should be made in sheet iron. Care must be taken that the oil from the bearings is not allowed to drop on to the marble being worked.

For moulding purposes, cutter blocks are made of chilled cast iron, steel, composite metal, brass, emery composition, and even of stone itself.

Where revolving cutter barrels are employed, in lieu of the ordinary forms of cutters used, a series of thick circular saws of small but varying diameters will be found extremely useful for roughing-out purposes, and can be so arranged on their barrel or spindle that mouldings of considerable depth and some degree of complexity may be roughed out with a minimum amount of "plucking" on the stone. The mouldings thus produced must, however, be finished by a fixed scraping cutter shaped to the desired outline, or by hand.

For cutting gains or grooves in stone, separate groups or gangs of chisels are usually employed; these chisels, which are pointed similar to Fig. 33, are worked by steam, and have a reciprocating movement. They are mounted in frames arranged on either side the main frame of the machine, which moves horizontally on rails over the stone to be channelled. The chisels or cutters work outside the rails carrying the machine, and the frame carrying them should be arranged to swivel, so that the cutters may be set to work at any desired angle. The machine should be fitted with varying rates of feed, which should be regulated in speed according to the nature of the stone being worked.

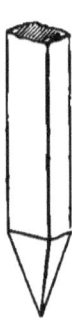

FIG. 33.

This form of machine is also employed in quarrying, and by making gains in the stone it can be more readily split off in layers.

Some years ago, Mr. George Hunter brought out a machine for cutting stone, &c., out of the living rock in the quarry itself; and, although the subject is slightly beyond the scope of these pages, the machine, from its ingenuity, is deserving of a further notice.

In this machine, the cutting tools, instead of being placed in a single row round the periphery of a disc, were fixed in rows of twos and threes alternately across the margin of the wheel-like disc, so as to clear away a wider space. The outer portion of this wheel-like disc was a ring of fine malleable cast iron, armed on the outside with tools, and carrying a cogwheel outline. Not to enter into small details of fittings, this cogwheel is made to revolve on a broad metal plate as its axle. This broad plate is of great strength, and forms four-fifths of the diameter of the entire cutter, and can be firmly bolted to the machinery frame by any part of its surface nearest to the cogged wheel which carries the tools, and the latter so held is made to revolve by a pinion around it. This arrangement allows eccentrically-held cutters to penetrate the rock to a depth exceeding the semidiameter of the disc. In the circular saw, with a central axle, the blade can only penetrate to so much of its semidiameter as is clear of the axle and collar, and great force would be required to hold such a cutter up to its work in the rock; but in the machine now before us, the cutter wheel is buried in the cut up to the point at which it is held, and, practically, allows of a cutter of 3ft. 4in. diameter burying itself to the depth of 2ft. 3in.; and as the cutter cuts out at a chord smaller than the diameter, the tendency of the out-coming tools is to draw the cutter into the cut, instead of forcing it out. A machine of this kind, cutting horizontally, works with great freedom, and advances rapidly through slate rock upon which it is employed. But when a cutting wheel on this principle is applied to make a vertical cut, a still smaller surface of the broad axle-plate is occupied by the holder, as it can in that position be grasped on both sides, and the axle carrying the pinion can be passed through the wheel and supported on double bearings.

The machinery, including the cutter wheel, for vertical

cutting, is fixed on a carriage running upon rails, and worked by a wire rope. The cutter wheel is gradually brought down from its travelling position by a worm and worm wheel to press upon the rock till it buries itself up to the holder, when it is fed forward by a self-working screw, attached by chain and swivel to some point in advance, or by winding directly upon a chain.

A tolerable level having been first obtained on the face of the quarry, and a line of rails pinned down in the direction of the cut to be made across the greatest length, the machine commences its advance, leaving a deep groove behind it 2in. or 3in. wide, and 2ft., 3ft., or 4ft. deep. A series of parallel cuts may afterwards be made, or two disc cutters on the same carriage frame may be advantageously used to make two cuts at a time. An opening at the commencement of the first cut end must then be got out by blasting or otherwise; but afterwards, if the rock has any sort of cleavage or layering, it may be wedged up from below. The rock from the first two cuts being thus removed, the vertical cutter may readily be applied to cross-cut the longitudinal grooves into squared blocks to be removed by under-wedging, or partial undercutting, when there is not a favourable cleavage. The principle of the undercutting machine is precisely the same as that of the vertical cutter. The cutter plate, however, has to be held horizontally by one side only. The cutter lies under the frame that carries it, and is advanced into its cut by worm and worm wheel, as already described. When buried up to its holder in the rock, the cutter traverses along a slide-frame 12ft. long. At the end of the 12ft. the slide and its carriage frame are pushed forward on the wheels for another length, and there fixed so as to leave the cutter free in its previous cut; it again proceeds on its journey, fed forward by its self-acting screw, and so on to the end of the opening.

CHAPTER XI.

SCULPTURING MACHINERY.

MANY attempts have been made to construct carving and sculpturing machines; these have met with a very small degree of success, and there still remains in this direction ample scope for inventive genius.

The best-known machinery is that patented by Mr. Jordan, of London, in 1845, and a variety of machines based on this patent have since been constructed. It was designed for carving and copying irregular forms in wood or stone. The material to be shaped, and the model, or "dummy," were fixed on a horizontal table running on wheels transversely on another table or frame, which was arranged to move in a longitudinal direction, so that by the straight line movement in two directions the table could be made to have a motion in every part of its own plane. The model and material to be shaped were made to swivel on centres, and so arranged that by means of a lever each could be turned simultaneously on its axis. The cutters were carried on a vertical slide this vertical slide was raised or lowered to the work, which was fixed on the travelling table beneath by means of a treadle. A tracer guide acting on the model produced, by the aid of the cutters, facsimiles in the piece or pieces of material.

Kennan, of Dublin, some years since, adapted a lathe to sculpturing purposes. The cutter, which was driven at a tolerably high rate of speed, and the guide, or tracing point, were mounted on a triangular steel bar, working on a universal pivot at one end, attached to the fixed head of the lathe, and suspended freely at the other end by a balance weight and cord over a pulley. The model and the block to be sculptured were fixed upon chucks mounted on two carriages, which were adjusted in position on the bed of the machine, according to the required proportional dimensions, when, of course, the relative distances of the model and the block from the pivot of the bar must correspond with the tracer and the cutter. The chucks were turned simultaneously on their centres by means of two worms on one rod, turned by means of a hand wheel at the end of the machine, and gearing into worm wheels on the chuck spindles, so that every part of a round object accompanied by the block may be brought before the tracer and the cutter respectively. The driving band was kept at a uniform tension by being passed over straining pulleys on a weighted lever at one side of the machine.

In another machine for copying sculpture, the model and the block are placed on two revolving tables, turned in unison by a worm shaft. With a dummy point, or tracer, at the model, as a guide, a small drill or cutter, turned by a cord passing over pulleys carried by a swing frame hung from above, operates upon the block. The point and the cutter are raised by a vertical screw in the frame. A ball-and-socket frame is applied for turning and undercutting, or to descend upon a horizontal surface. For cutting basso-relievos, a parallel traverse on two guides is employed with the screw.

The great engineer Watt also constructed a sculpturing

machine, which is said to have copied busts with some success. This machine has recently been fully illustrated and described.* It appears Watt constructed two machines, one for reproducing busts or bas-reliefs of the same size as the original, and the other one for making a copy of a reduced size. The first machine Watt called an "Eidograph," and it consisted, firstly, of an ordinary lathe, with treadle and fly-wheel, to supply the motive power; and secondly, of two tall uprights about seven feet high, carrying at the top a slide on a strong horizontal bar, the slide being capable of motion horizontally, either at a slow or quick speed. Then, hinged to this slide is a light square frame of metal, and, at the outer edge of this, another light square frame of metal is hinged, so that the lower edge of such frame is capable of motion up and down, or in and out, like an elbow joint, and horizontally when the top slide is moved. The weight of these frames is balanced by levers and balance weights and chains above, and the lower edge of the second frame is furnished with a "feeler" or "guide," to traverse over the original model, and a drill driven at a high speed by a light cord to cut the work or copy; so that by handling the feeler carefully, and tracing over the original in all directions, a piece of marble or alabaster or wood, placed in the machine alongside of the original is cut to a perfect copy by the machine without fear of any mistake, and without any special skill on the part of the operator.

The slow motion to the slide above, carrying the frames and feeler and drill, is worked by a convenient handle and tangent-screw when cutting, and the quick motion can be thrown into gear with the lathe wheel to run back. The quick motion has a coarse traversing screw, having a nut

* Proceedings of Inst. Mech. Engineers, Nov., 1883.—Paper by E. A. Cowper.

in halves, that can be closed or opened; and the slow motion has a fine-threaded screw with a similar nut, so that it also can be thrown into gear or released. There is a noticeable feature in the frames above mentioned, and that is, that in order to prevent their springing or going "winding," they are practically formed into "solids" by the erection of the outlines of a pyramid on each; this plan gives extreme stiffness at the expense of very little weight. Specimens of the work done on this machine are in existence, and both the original and copy can be mounted in their places on the machine, and be turned precisely together by a pinion gearing into the two wheels on the mandrels of the carriages on which the articles are placed, so that undercutting could be properly accomplished, as well as straight cutting into the work by the drills. The drills, circular cutters, and other cutting tools are excellent, some being formed for roughing-out apparently, and made to cut in steps, some in the forms of globes with the whole surface formed into numerous cutting edges, so that it was a cutting globe so to speak, and could go anywhere, as it could cut in any direction. Watt called his second machine for making reduced copies a "Diminishing Machine." The machine consists, firstly, of a lathe bed, with fly-wheel and treadle for obtaining the motive power for driving the drill; secondly, of a stout hollow tube, forming a long lever, fulcrumed at one end on a universal joint, so that the other end can be moved in any direction about the centre. This lever carries a "feeler" or blunt point near its outer end, and a drill near the fulcrum, so that whatever motion the feeler has, the drill has (say) one eighth part as much. The lever is balanced. The slides above named slide on the bed of the lathe, and are moved by a pentagraph or arrangement of levers, to give one eighth as much motion to the work to be cut, as to the original, so that every dimension shall be in proportion. A further motion is provided for turning round the original and the copy, as is sometimes necessary when undercutting a bas-relief, and, of course, when copying the round figure.

In the London International Exhibition of 1862, Messrs. Cox & Son, of London, exhibited a machine for carving in wood; applicable also for carving in stone and marble. It was founded on Jordan's well-known patent; and by means of revolving cutters and a tracer-guide, it copied from a model in duplicate and triplicate. In place of the tracer and cutter moving over the model and the work, as in some other machines of this class, they were stationary in a balanced frame, with freedom for vertical motion only. The vertical spindles are reported to have made from 5,000 to 7,000 revolutions per minute—this speed, however, was for working wood—whilst the model and the work were traversed under the tracer and cutter on a table capable of horizontal movement transversely and longitudinally. This machine was said to be capable of roughing-out four panels in four hours, which were finished by one man in a day, and which would take a man four days to carve by hand.

We believe this or a duplicate machine was worked with success for some years in the Belvedere-road, Lambeth; but we are unaware whether it is at present in use.

Many other attempts have been made to utilise the Blanchard copying lathe, &c., for sculpturing purposes, but without any permanent commercial success. Although there may be considerable difference of opinion—in an art sense at any rate—as to the advisability of reducing sculpture, when possible, to a mechanical process, much can be said on the other side as to the gain occasioned, in the art education of the masses, by the reproduction by mechanical means of standard examples at a nominal cost; or, at any rate, by allowing machinery to do all the rough work, leaving the artist or art workman to give the finishing touches. Before, however, even this can be accomplished successfully, great improvement in the already-existing machinery must be made.

CHAPTER XII.

POLISHING BEDS.

These have undergone comparatively little alteration for many years. The polishers are usually actuated by a crank from the main or an intermediate shaft: this, by means of a connecting rod, sets in motion a swing frame acting as a pendulum; from the base of this swing frame rods run horizontally to the rubbers, or polishers, which work to and fro on a bench or table. The swing frame is usually suspended on centres fixed to beams above the machine, and the rubbers can be adjusted to the width of the marble or slate by moving their rods transversely on the bottom rod of the swing frame.

Whitworth describes an American polishing machine in which the stone is polished by a flat circular disc of soft iron, which is made to revolve horizontally. The axis of a disc is fixed at the end of a heavy frame, which moves round a strong centre shaft in a radius of about 12ft. The polishing disc revolved at 180 revolutions per minute; it was driven by a strap, to which motion was given by a driving pulley, fixed on the centre shaft; the disc was guided and its pressure regulated by hand. It was represented as being capable of polishing about 400 square feet of surface in a day of ten hours. Our illus-

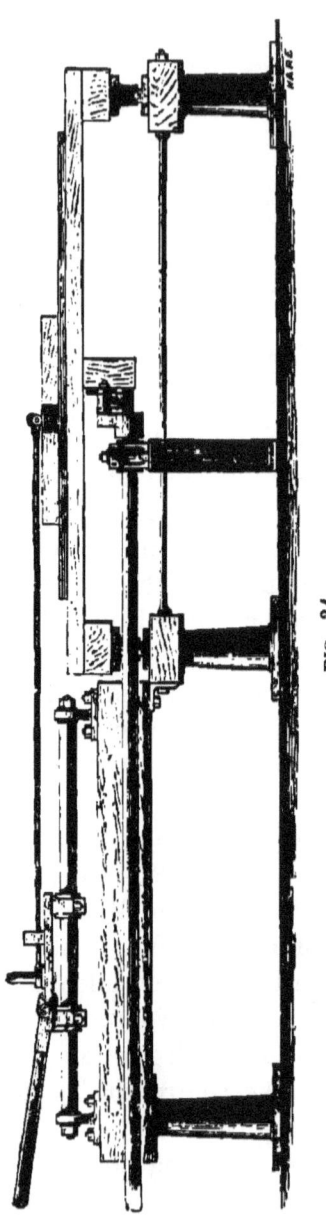

Fig. 34.

tration, Fig. 34, represents a horizontal polishing machine for marble, granite, stone, &c. It will be seen from the sketch that the bed on which the stone to be polished is placed, receives by means of a crank a horizontal motion at right angles to the travel of the polishing blocks: this arrangement effectually prevents lines being formed on the face of the stone.

The chief grinding and polishing materials are given by Knight as follows:—(1) Abrasive substances used in the solid form:—Grindstone, hone, oilstone, charcoal, emery composition, fish-skin; (2) abrasive substances used in powder, stated in about the order of their hardness:—Diamond, sapphire, ruby, corundum, emery, sand, flint, glass, tripoli, Turkey-stone dust, rottenstone, slate, pumice, chalk, oxide of iron, colcothar, crocus or rouge, oxide of tin or putty powder.

The abrasive powders are applied by circular discs, or

sheets of iron, which cause them to act as saws. On the periphery of wheels, which act as grindstones, glazers, or buffs, according to the quality of the material and the terms of the trade; on the plane surface of discs which form laps, on the ends of rods which act as drills, and on slips of wood which act as files. They are also spread upon cloth, paper, leather, &c.

For the guidance of the uninitiated, we may say a grinding or polishing lap is a wheel or disc, or a piece of soft metal, used to hold cutting or polishing powder. They are made of brass, cast iron, copper, lead, and various alloys. Laps are usually made to revolve on a vertical or horizontal spindle; but their arrangement, of course, varies with the nature of the operation carried on.

In polishing marble, &c., sharp fine sand, pumice, or snake stone is first employed, afterwards a finer, and sometimes a third; after the finest sand is used, emery of different grades is employed. Tripoli powder is then used, and the polishing process is completed with putty powder (oxide of tin). The polishing plate for sand is made of iron, and its size, length of stroke, &c., is varied according to the area of the material being polished. A polishing plate made of an alloy of lead and tin is generally employed with the emery, and for the finishing process old linen cloths and rags forced tightly into an iron frame. A small but constant supply of water is also necessary.

The polishing process is, however, varied according to the nature of the work and the degree of finish required; in working dark marbles up to a fine surface crocus is often used before the oxide of tin is applied. It is important that the marble is carefully washed after each separate polishing material is used; as, should the different particles become mixed, the work would in many cases become considerably damaged by scratches. It is important also that the sand employed should be of uniform quality and the particles of equal hardness; as, should some be much harder than others, deep scratches will be the result. To secure this uniformity the sand should be levigated and washed, and the harder particles, being usually of a greater specific gravity than the rest, may be separated with tolerable facility. Washing the sand will, in practice, be found much preferable to sifting.

For rubbing or polishing blocks of moderate dimensions, a modification of a wall drilling machine may be used with advantage. As will be seen from the sketch (Fig. 35), to the end of the vertical spindle is attached a flexible or universal jointed shaft, which is under the immediate control of the workman, who can regulate the pressure and position of the polishing disc, which is fitted to the bottom end of the shaft.

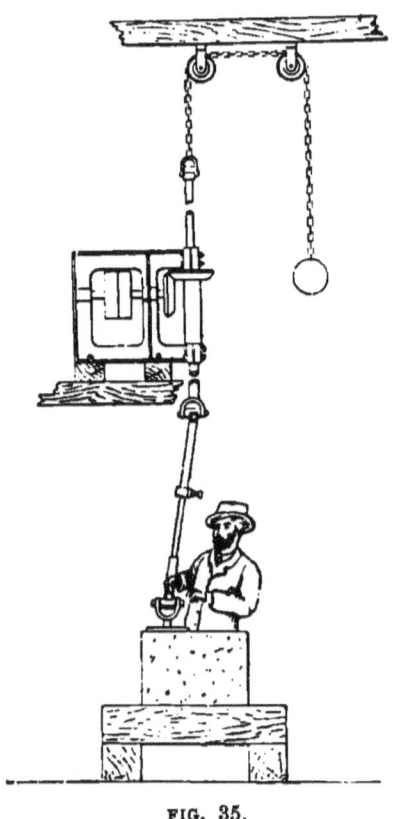

FIG. 35.

The sliding bracket which carries the vertical spindle can be raised or lowered at will, it being suspended and counterbalanced by a chain and weights passing over pulleys attached to the ceiling. Motion is given to the spindle by a pair of bevel wheels. The table carrying the stone is, in the sketch, a fixture; but where large quantities of heavy stones are polished, a series of these machines should be mounted and a travelling carriage, carrying several blocks of stone at once, made to traverse on rails beneath the polishing discs. In this case the discs are mounted on the ends of the vertical spindles, and the flexible shafts are dispensed with. Motion can be given to the first disc spindle by a pair of bevel wheels in the usual way, and communicated to the other spindles by intermediate toothed gearing.

For polishing small slabs, a variety of hand-rubbers are in use, one of which we illustrate (Fig. 36). The

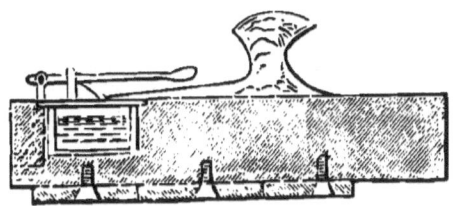

FIG. 36.

polishing material is carried in the chamber formed in the rubber-block, and is allowed to escape, as required, through a hole in the front of the block, by touching the trigger shown in the sketch, which opens and closes a valve. The polishing pad usually consists of folds of linen, but in Italy a slab of lead is used instead of linen. Several patents for marble-polishing machines have been taken out, including one by Maloy. In this machine the marble slab is placed on a slide, and polished by means of an endless belt, which is kept tight by an adjustable "idler" pulley. Another plan for grinding and polishing marble is to mount a long grinding cylinder on a spindle, and to give it a combined rotary and reciprocating movement, the slab of marble to be acted on being placed on a table beneath.

In grinding and polishing lithographic stones, a slow compound angular and reciprocating motion is required.

CHAPTER XIII.

TURNING LATHES.

MARBLE, granite, and many kinds of stone can be turned in a lathe with tolerable facility. Stones of a moderate degree of hardness can be turned by a piece of square cast steel drawn down to a point. We need hardly say, however, the steel employed should be of a high grade. For turning granite, rolling cutters are now generally employed. These are made of hard cast steel, and their dead contact being small, the friction or grinding action on, and consequent wear of, the cutters is largely reduced. After the object is roughed out, the tool marks are removed or ground out, usually with a series of sandstone, snake stone, pumice stone, and other materials, which are graduated in fineness, till all the marks and scratches are obliterated. For polishing purposes flour of emery, putty powder (white oxide of tin), and other substances are applied by means of a lap or a bundle of cloth.

For cutting circular slabs, &c., revolving cylindrical cutters of wrought iron, constructed on a similar principle to the crown saw and mounted on a disc, can be used. This is usually fitted on an adjustable vertical spindle, somewhat similar to that used in a recessing machine for working wood. For cutting large circles the cutters can be mounted at the periphery of a large disc, or at the end of cross arms.

For turning granite, marble, and very hard stone, Brunton and Trier's revolving cutters are the best that we are acquainted with, as, owing to the great reduction in friction accruing from their use, the number of revolutions of the stone can be largely increased; consequently, the amount of work turned out is much in excess of that produced by the ordinary steel tool acting with a dead pressure on the stone. We believe the first application of the principle of revolving cutters was to the turning of stone, and the cutter being once set in motion by contact with the stone continued rolling, and being placed at an angle of about 25° to the axis of the stone chipped the surface in a spiral line, as the slide rest and tool holder moved along the bed of the lathe.* It was found that the concurrent revolutions of the stone and the cutter reduced the attrition largely, and a considerable speed of surface rotation was therefore attainable. The stone, before being placed in the lathe, should be dressed to a rough octagon shape; and where taper columns are produced, the lathe should be fitted with double expanding beds. These revolving cutters are made of steel for working granite and hard limestone; but for gritstones, sandstones, and freestones chilled cast iron answers well. The steel cutters are made either in the form of flat discs or cones; the cast-iron cutters always in the form of cones. The advantages of the conical form are that the edge is always on the hardened or chilled surface, and, in the case of cast iron, the rest of the thickness of this cutter is comparatively soft, and therefore more quickly ground. The diameter of the cutter also is much less reduced by wear than is the case with disc cutters. A

* I. M. E., January, 1881.

simple and efficient plan of fixing these cutters on their chuck-spindle has been devised: a split nut is used, the conical part of which enters into a corresponding hole in the cutter. When screwed up, the nut grips tightly the thread of the spindle.

The advantages claimed by the inventors for this system of rolling cutters for turning purposes are:—(1) That the friction between cutter and stone being absent, no water is necessary to keep the temper of the tool, thus preventing the effect water has in opening the soft veins of marble, &c. (2) Absence of dust. (3) Greatly diminished pressure on the column, even whilst taking much heavier cuts than is possible with the ordinary tool. (4) Complete absence of "plucks," a perfectly regular and, therefore, quickly polished surface. (5) Granite columns perfectly round and true, and without holes, as is often the case in columns masoned by hand. We ourselves have seen these cutters turning Cornish granite, marble, &c., and they must, we think, be held to be a decided step in advance over the old system of stone turning.

In turning stone it is important that it is truly and firmly held in the lathe; for turning the smaller articles in marble, &c., a chuck, consisting essentially of an enlarged centring head, is often employed; this receives and retains the article to be turned. The centring head is provided with a centring "spud," and an open-ended adjusting cap, the latter constructed so as to be longitudinally adjustable, and of such form as to securely hold the article against lateral displacement. The barrel of the head is screw-threaded on its outer surface, and the cap is constructed to hold the article at a point outside the centring head.

CHAPTER XIV.

STONE-BREAKING MACHINERY.

ALTHOUGH stonebreakers can hardly be considered to be stone-working machines, their products are largely used in roadmaking and similiar contractor's work, and it may not be out of place here to notice them briefly.

Machinery for breaking stone possesses an immense advantage over hand labour, and the stones are more uniform in size. The stone is usually broken by means of jaws of chilled cast iron, to which a kind of rocking, knapping, or reciprocating motion is given. In the best types of machines, when road metal or stone is required broken to a certain size, what are known as cubing jaws are fitted; or when a fine material is wanted, the cubing jaws are replaced by crushing jaws. In addition to breaking stone, this class of machines may be used with advantage in crushing or breaking hard rocks, ores, fossils, pyrites, flints, coprolites, limestone, emery, phosphates, fireclay, coal, and other materials. In some cases the machine may be employed with advantage in the quarry itself, the broken material being removed by means of an elevator. In the earlier forms of crushing machines the movable crushing jaw, when forced forward, was made to compress a spiral spring imbedded in indiarubber; this withdrew the crushing jaw after it had completed its stroke. This plan, although acting tolerably well, consumed a considerable amount of power. A simpler and better plan for performing the same operation has, however, latterly been introduced. This consists of an arrangement of levers which are adjusted as required. We have recently seen working a multiple-action stonebreaker, of somewhat novel construction. In this machine the crushing jaws are divided, either half being worked by a separate eccentric attached to a main eccentric shaft, each eccentric operating a separate connecting-rod and set of toggles; they are also arranged to balance each other whilst in motion, thus reducing the strain on the machine and consequent power required to drive. One jaw only operates on the stone at the same time; but the double connecting-rods employed travel through twice the space of the single rods employed on most machines. The crushing jaws are withdrawn after each stroke by operating one on the other by means of coupling the rods of the jaws by a cross lever mounted on a stud at the end of the machine. The main framings of this machine are of lighter construction than is usually employed, the inventor claiming that, through the balancing of the moving parts, excess of metal is not required. Another plan recently introduced is to arrange a rocking lever vertically on a fixed centre at its lower end, the rocking motion necessary for crushing being obtained by an eccentric operating on the other end.

A movable crushing jaw is arranged on either side of the lever, and connected with it by loose toggles, kept in position by an adjustable rod. When in work, the rocking motion of the lever imparts an alternate motion to the crushing jaws, and by adjusting the jaws two different sizes of stone can be broken at the same time.

For crushing road metal, a very simple arrangement—

and one which does away with the revolving screen usually employed—is to combine a lever jaw and a chilled cast-iron roller with corrugations at right angles (Archer's patent).

In the construction of stonebreakers, strength, with simplicity of construction or renewal, are the points to be aimed at. The cubing or crushing jaws should be made adjustable and reversible. Chilled cast iron is the best material to employ; for use in remote districts, loose or adjustable jaw faces should also be fitted. The shaft giving motion to the crushing jaws should be of steel, as frequently a large amount of strain is put on them; and if not of good material and of considerable strength, they may be either bent or broken—rather an awkward circumstance should the nearest engineer's repairing shop be some hundreds of miles away. As the friction on the bearings is also great, especial attention should be given both to their construction and lubrication.

Our illustration, Fig. 37, represents one of Blake's patent stonebreakers, by H. R. Marsden. As will be seen from the illustration, it is of massive construction, to overcome the great strain and vibration in working. It is fitted with reversible cubing jaws, made in sections and with faced backs. It is arranged with steel-bar toggle bearings and improved grooved expanding toggles. The sizes of the cubes of stone can be readily regulated, and the jaws set to break as small as the finest gravel if required. A modification of this machine is also constructed for grinding cement, limestone, silica, shale, flints, &c., to a powder; and by combining an elevating and sieving apparatus with it, a very considerable degree of fineness may be obtained. The illustration shows a side view or elevation (sectional). A is the main frame; F is the fly-wheel shaft, which should make about 250 revolutions per minute, and by means of the belt from the engine on to pulley O. The larger circle inclosing F shows the eccentric. H is the connecting-rod, which

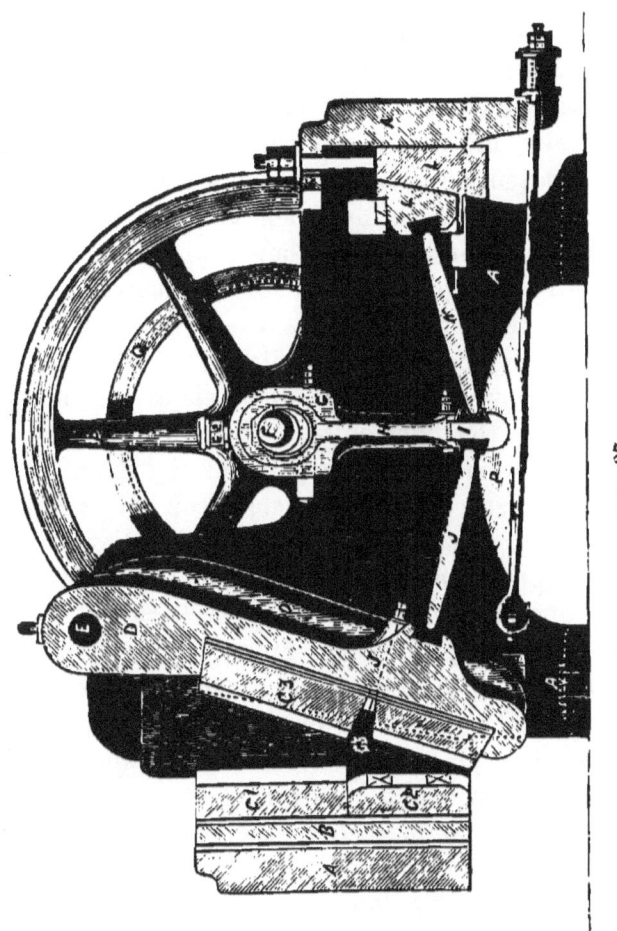

FIG. 37.

connects the eccentric with the toggle plates, J and K, which have their bearings forming an elbow or toggle joint. These bearings, it will be seen, are all removable; and the same may be said of all the pillows and bearings throughout the machine. C is the cutter for taking up the wear of the pillow or bush in which the eccentric works in the connecting-rod head; I is the cotter for keeping securely in their places the bearings of the toggle plates. B is a false back, accurately planed on the surface and bedded to the frame. A, C, 1, 2, 3 and 4 are jaw-faces, which are fitted with patent metal strips on their backs, and thus find a firm and even bed for themselves, preventing sudden strains which they necessarily have to encounter, causing them to break and give way; E is the shaft of the spring-jaw, which is cottered to it and itself rests upon each side of the main frame A; J is the key for keeping in position the spring-jaw faces C, 3 and 4; D is the spring-jaw itself; C, 5 is a cheek, of which there are two, one on each side of the mouth, and which keep the fixed jaw in position and prevent the stone from wearing into the sides of the main frame; P is the fly-wheel; L is the toggle block and wedge. The toggle block, as will be seen, takes the cushion or bearing of the toggle plate K, and itself is held in position at one height, always by means of lugs on each side of the frame; whereas the wedge L moves up and down at the will of the user of the machine, thus reducing or increasing the opening of the jaw at the bottom, and so regulating the size of the product. The hook bar underneath the machine has an indiarubber spring attached to it at the end, which is compressed by the forward movement of the jaw, and aids its return. Every revolution of the eccentric causes the lower end of the spring-jaw D to advance toward the fixed jaw about $\frac{1}{2}$in. and return; hence when a stone is dropped in between the jaws, it is broken by the next succeeding bite; these fragments then fall lower down and are broken down, until they are small enough to fall out at the bottom. The distance between the jaws at the bottom limits the size of the product, and these distances, as before named, can be regulated at pleasure by turning the screw nut which raises or lowers the wedge L. Greater variation may also be made by substituting the toggle J, another longer or shorter, extra toggles of different lengths being furnished for this purpose.

CHAPTER XV

SLATE-WORKING MACHINERY.

Before we notice machines used for quarrying slate, it would be well to consider briefly the peculiar geological features of its formation. It is usually found in large beds divided by fissures or joints of other substances; but it differs from ordinary stones by being capable of being split into thin sheets or plates. The direction of these laminae in the slate always lies either vertically or obliquely to the plane of the bed, and never parallel with it. The object to be borne in mind, whether quarrying by hand or machine, is how to cut out or detach suitable blocks of slate with the least amount of waste and labour.

In quarries where there are many "faults" or "dykes," of greenstone, &c., and where the cleavage is interrupted, machine quarrying is not very applicable, as the results obtained are not in proportion to the outlay for machinery, &c.; but in large quarries of tolerably uniform strata, there is little doubt a well-constructed machine devoid of complexity can be worked with very economical results. There are several varieties of slate found, including talcous, mica, flinty, and clay. The last is the one chiefly used; it varies considerably, however, in quality in different districts, but is generally found suitable for machine conversion in some form or other, except where it is soft or rotten in texture, or where it contains much felspar or hard rock.

Of late years slate has, in addition to roofing purposes, been largely employed in the manufacture of mantelpieces, billiard tables, cisterns, &c.; and slate, where the cleavage is bad, may in many cases be used with advantage for these purposes. After the slate is quarried, if suitable for roofing or writing slates, it is split and sorted according to its quality or size. If the block is better suited to other purposes, it is converted into suitable slabs by horizontal reciprocating or circular saws; it is afterwards faced on a rubbing-bed, and moulded if required. When the slates are used for roofing purposes, they are first of all split to the desired thickness by means of a wedge and mallet, or in some cases by a machine. This is usually done as soon as the slate leaves the quarry, as it becomes difficult, and in some cases impossible, to split the slate into leaves should the quarry-damp or water be dried out of it through exposure to the sun and air. A plan often practised to see if a slate is of fine quality is to heat it very hot in a fire and plunge it into cold water : if it will stand the test without fracture, its quality may generally be relied on. For squaring the edges of roofing or other slates, machines are employed, worked either by hand or power. With the hand-power machine the slate is laid on a table and trimmed by means of hinged knives worked by levers. These knives are sometimes made curved; but, in any case, whether worked by hand or steam, they should be arranged so as to give a kind of shearing cut to the slate, which renders it less liable to break than if the whole width of the blade acted on the slate at the same time. In machines worked by power, to prevent the breakage of the slates, they are usually held between spring cushions, which receive the jar or vibration caused by the knife or knives striking the slate; the knives in this case are arranged to act after the fashion of a guillotine. A machine used considerably for dressing slates in the Welsh quarries is Francis' patent, known as the "sword arm," made by De Winton & Co., Carnarvon. This consists briefly of an imitation by mechanical means of the elastic shearing cut of the ordinary dressing tool. This is accomplished by mounting one end of the "sword arm" upon a rocking shaft moving in suitable bearings. Elasticity in the cut is obtained by suspending the arm by a spiral spring at about one-third of its length from the rocking shaft. Motion is given to the arm either by a foot-treadle movement or other motive power. This machine is of the simplest construction, and is rapid and clean in its cutting. At a trial of its capabilities some time since, a machine worked by a treadle finished 207 slates, of various sizes, of a total superficial area of 358ft., in twenty-five minutes, a somewhat remarkable feat.

Another machine for dressing slates, but driven by steam or other power, is Gream's Patent. It consists briefly of a cast-iron frame carrying a spindle, on one end of which is fitted a fly-wheel and driving pulleys. On the spindle between the sides of the frame are keyed two cast-iron rings or hoops, to which are bolted opposite to each other two curved knives made of steel. These knives are made to revolve, and when in motion pass close to a steel cutting edge fixed on the top of the framing on which the slate to be dressed is held, and the knife coming round shears it through with a scissors-like cut. An adjustable fence is fitted for instantaneously gauging the sizes; at the same time it serves as a stop to hold one cut side against when dressing the next side, thus insuring the slates to be perfectly square. The driving pulleys make about 80 revolutions per minute. It is important that the fly-wheel, &c., in this machine is accurately balanced, so that the strokes of the knives are even and without jump. Should the wheel be out of balance, and any centrifugal force set up, the slates being dressed are much more liable to fracture.

Amongst other machines designed for cutting and trimming the edges of slates, may be mentioned one introduced at the Maen Offeren slate quarries, Festiniog, some years since. This consisted of a fly-wheel working on a horizontal axis, with a number of knives for trimming the slates fixed at equal distances on the sides of the wheel; or, if two slate cutters worked at the same wheel, knives were arranged on the other side of the wheel in alternate order. The knives were fixed at such distances from the spokes of the wheel as to admit the slate being presented to the knife without the projecting end coming into contact with the spokes of the wheel. The knives were arranged radially from the axis of motion, so that the inner end of the knife struck the slate first. The slate to be cut was rested obliquely on a cutting edge fixed on the framework of the machine, and received the revolving knives progressively from its inner to its outer extremity, it thus giving a shearing cut.

For edging and trimming slabs of slate, circular sawing machines of the class already described for sawing stone are employed, the operations of rubbing, polishing, and horizontal or straight sawing are also practically the same. For rounding the edges of slabs, an old machine known as the "ridge roll" is still largely employed.

Amongst the most advanced machines for sawing slate is De Winton's Patent Hydraulic Feed Circular Saw. In this machine the spindle carrying the circular saw is driven by pulley and belt in the usual manner. The patent consists in the arrangement for giving independent travel or motion by hydraulic pressure to the table which carries the slabs or blocks to be sawn. The table is connected on its underside by a bracket and nut to the piston-rod head of an hydraulic cylinder, which is bolted underneath the table to the framing. A supply pipe leads to both ends of this cylinder from a valve box, which is connected with a main pipe having a pressure of about 60lb. per square inch.

On shifting a slide valve in the valve-box by means of a lever, the water pressure is admitted to the front or back end of the cylinder as the case may be, and the piston is then driven along carrying the table with it. Whichever end of the cylinder is under pressure the pipe from the opposite end serves as an exhaust to carry away the water which filled that end of the cylinder at the previous stroke, and it is by regulating the area of the exhaust outlet on the valve-box by a screw valve that the speed of the travelling table is controlled, and can be varied and regulated as required. A fast or slow movement can thus be given to the table, according to the thickness or hardness of the slate being cut, and quick return motion of the table, preparatory to making a fresh cut, may readily be obtained. The makers claim that a very small quantity only of water is required to work this feed, a 3in. main pipe being sufficient to work twelve pairs of saws. This form of feed motion possesses the advantage of being very steady and even in its movements, thus preventing the saw from being forced, or the slate jagged and marked, which is occasionally the case with chain or rack feed.

So great is the demand for school slates, which are made from a slate of a fine and soft quality, that a variety of very ingenious machines have been invented, with the object of lessening the cost of their manufacture. One of these is a modification of a planing machine for wood, adapted for planing and dressing the slate frames. The frames are fed forward through the machine and brought under the action of horizontal and vertical cutters, which plane the sides and edges and round off the corners. One machine for grinding and polishing school slates, used considerably in America, possesses some novel features of arrangement, which may be described as follows:—The slates are supported on carriages mounted on wheels or rollers. These carriages are connected by arms with a vertical shaft, by which they are caused to move slowly round in a circular track, arranged with a series of inclines. The distance apart of the corresponding parts of each similar incline is exactly equal to that between the front and hind wheels of the carriage, which is thus caused to remain perfectly level, as it rises to present the surface of the slate to the grindstones, which are arranged above them, and falls during its forward movement. The manufacture of slate pencils possesses also some features of interest. Knight briefly describes one process as follows:—The slate is first split into slabs 1in. to 2in. thick, which are then sawn into blocks 6in. to 7in. long, and 4in. or 5in. wide. With a thin blade of steel and a hammer these are split into plates about one-third of an inch thick, which are next passed between two flat-edged knives to plane them. The plate is then fed to a machine, in which it is passed successively beneath a series of grooved cutters, each of which cuts a row of deeper incisions into the slab, until, on emerging from the machine, its upper side is covered with convex flutings, the channels between which penetrate half through the stone. It is then transferred to a second machine, where its other side is subjected to the action of a series of similar cutters, by which the pencils are completely rounded and separated from each other. They are next sawed to uniform lengths, the sizes varying from 3½in. to 6in., and, finally, pointed on a grindstone. In some cases they are afterwards painted.

The dust and waste, which is said to amount to 90 per cent. of the original material, is utilised by grinding to an impalpable powder, which is used for mixing with paper-pulp to give it body and enable it to receive a satin surface. As the stone contains over 30 per cent. of alumina, the refuse is also available for the manufacture of alum.

Slate pencils are also made from quarry waste; the process of manufacture may be thus described:—Broken slate from the quarries is put into a mortar driven by steam, and pounded into small particles. Thence it goes into the hopper of a mill, which runs it into a bolting machine, such as is used in flour mills, where it is bolted, the fine, almost impalpable flour that results being taken into a mixing tub, where a small quantity of steatite flour, manufactured in a similar manner, is added, and the whole is then made into a stiff dough. This dough is thoroughly kneaded by passing it several times between iron rollers. Thence it is carried to a table where it is made into charges—that is, short cylinders, four or five inches thick, and containing from eight to ten pounds each. Four of these are placed in a strong iron chamber, or retort with a changeable nozzle, so as to regulate the size of the pencil, and subjected to severe hydraulic pressure, under which the combination is pushed through the nozzle, in a long cord, like a slender snake sliding out of a hole, and passes over a sloping table, slit at right angles with the cords to give passage, with a knife which cuts them into lengths. They are then laid on boards to dry, and after a few hours are removed to sheets of corrugated zinc, the corrugations serving to prevent the pencils from warping during the process of baking, to which they are next subjected in a kiln, into which superheated steam is introduced in pipes, the temperature being regulated according to the requirements of the articles exposed to its influence. From the kiln the articles go to the finishing and packing room, where the ends are thrust for a second under rapidly-revolving emery wheels, and withdrawn neatly and smoothly pointed ready for use. They are then packed in pasteboard boxes, each containing 100 pencils, and these boxes in turn are packed for shipment in wooden boxes containing 100 each, or 10,000 pencils in a shipping box.

In planing slate, a modification of the ordinary planing machine for iron is usually employed. In lieu of the ordinary tool for cutting iron a wide sheet-steel cutter is used; this is mounted in an adjustable slide and used in the ordinary way, the slate to be planed being traversed to and fro beneath it.

For holing slates, a machine, arranged with one or more punches, is used. These are usually brought down to the slates by a lever acting directly on to a quick-threaded screw, or by a lever acting on a horizontal spindle, mounted in a slide, and carrying two pairs of bevel wheels and pinions acting on screws, which give a combined rotary and downward motion to the punches or drills.

Many attempts, more or less successful, have been made to cut out slate, stone, or marble from its natural position in the quarry or mountain. One of the latest is that of Mr. George Hunter, who has recently (1882) patented improvements in machinery for tunnelling and quarrying slate, and the methods of employing it, so as to quarry slate cheaply and rapidly, and obtain it in pieces of form and size suitable for being worked into commercial slates or slabs.

In constructing the tunnelling machine, the inventor employs as framing two three-armed end frames, connected together by a central tube. The arms of the frames are provided at their ends with setting screws, which can be screwed against the interior surface of the part of the tunnel already bored, so as to secure the machine firmly in position. Two of the arms of each end frame, which extend obliquely downwards, have mounted on them wheels or rollers, on which the machine can be run along the floor of the tunnel, when the setting screws are released. Through the central tube passes a strong shaft, which carries at its end a two-armed boring head, each arm having at its outer end a circularly-curved plate, at the front edge of which are fixed the cutters. The shaft is caused to revolve slowly by a worm working a worm wheel on the shaft, the worm being itself on a cross shaft mounted in bearings in the upper part of the frame. This cross shaft, or a shaft geared to it or connected to it by a strap, may be driven by bands from a distant motor, or by a compressed-air engine mounted on the framing of the machine. By the arrangement of three-armed framing having the lower arms spreading out, and having all the gearing above the central shaft, the inventor claims that great convenience is secured for the removal of the core of the boring after it is detached from the rock. For feeding the boring head onwards, the inventor applies at the rear end of the central shaft a screw, driven at a differential speed or kept at rest while the shaft revolves; gearing is arranged to reverse this screw so as to withdraw the boring head rapidly when it has cut its full length.

The undercutting machine is arranged so that it can cut either above or below, or both above and below, the mass of rock to be removed. For this purpose a vertical shaft is mounted in a framing, and caused to revolve by worm gearing, which may be worked in a similar manner to the tunnelling machine already described.

This vertical shaft has at either of its ends, or at both ends, a revolving saw armed with cutters at its edge, either saw being removable and capable of being slid along the shaft, so that the distance between the saws when both are used can be varied as required.

The frame is fitted to slide along two horizontal bars secured to girders, which are fixed in the working, and a feed-screw is provided to cause it to slide slowly along the bars, according as the cut affected by the saw advances. The frame is arranged so that the bars can be inclined, and the inventor thus inclines them for beginning the cut, the saw entering the rock obliquely from a part already opened until it has descended in an inclined line to the required depth, whereupon the bars can be set level, and the cut continued in a horizontal direction. When the saw-shaft and gear have travelled along the bars to their end, the screws that fix the framing are released, and while the saws remain stationary in the rock, the framing is advanced its own length, and when it is fixed again the cut is continued. By arranging the lines of the cut so as to cross the cleavage of the slate, blocks can readily be separated by the chisel from the body of the rock; and when two saws are employed together, or when one saw is employed under a floor already laid bare, it is advantageous to make the space between the saws or the depth of cut under the floor correspond with a dimension of commercial slates or slabs, so that the blocks separated can be cleft without material waste.

In cases where a floor is already cleared, the saw spindle may be driven by a wheel of considerable size, as it may partly extend over the cleared floor while the saw is cutting at the desired depth below the floor. The tunnelling machine can be used to drive a boring into the body of the slate, and then, after clearing a space, it can be turned into an attitude to drive other borings at angles to this, so as to suit the character of the stratification, and the saws can then be employed to cut into the exposed faces, leaving pillars, where necessary, to support the roof. Sometimes, in order to separate a mass along which saw cuts have been run, the inventor employs the following method:—A bar, made in halves, has each half hollowed, so that when the two halves are put together they make an internal cavity, or hollow; this hollow is charged with explosive, and the bar inserted along the bottom of the saw-cut, so that when the explosive is fired the mass receives from the explosion a blow which separates it from the body of the rock. The general arrangement of Mr. Hunter's machine is ingenious and practical, and as slate quarries become exhausted and the necessity for the more economical working of those remaining are essential, this and kindred machines should, and probably will, be more largely introduced, as there can be but little doubt that the greater proportion of the quarries are at present worked in an extremely wasteful manner. At the same time, by means of a well-arranged quarrying machine the slate or stone can be got in a much cleaner and more marketable form, thus increasing its value in the rough, and it subsequently requires less dressing.

CHAPTER XVI.

MISCELLANEOUS MACHINERY FOR WORKING STONE.

A CONSIDERABLE variety of special or miscellaneous machines for working stone have been constructed to suit the special requirements of a business or process. Amongst sawing machines may be mentioned an endless band rope arrangement. The rope is of steel, and passes over pulleys; it has a vertical motion, and combined with it a twisting one, which allows the strands of steel wire to cut their way into soft stone, it is said with considerable speed.

Amongst other machines, Mr. Hunter, senr., invented a chairing and boring machine for railway blocks; this was employed for facing the seat for the chair and boring the trenail holes.

Amongst quarrying machines may be mentioned one of American origin, especially adapted for cutting marble. In this machine a chain saw, armed with diamonds, is used. The diamonds are held in split screw bolts or clamps, which are tapered and forced into tapering cavities, causing them to contract upon the diamonds. The diamonds are applied on the edge of a rotating disc or the links of an endless chain, and the supporting standard of the disc or chain is so connected to the driving mechanism that it may be fed in either direction. A small steam engine is mounted on a travelling carriage, and, fitted with suitable intermediate gear, gives motion to the chain; and arrangements are made to feed the saw to its work and vary the direction of the cut as may be required.

A variety of machines for dressing mill-stones are in use; in one of these diamond points, mounted on discs, are used. The discs are made to revolve rapidly, and, the diamonds being brought into contact with the face of the stone, parallel grooves are cut in a similar manner to hand tooling. A guide bar is fitted so that the stone may be dressed right or left handed as may be desired. In another machine, the radial grooves in the stone are cut by means of a tool raised and dropped by a cam and advanced automatically along a radial arm attached to a central axis. In a hand machine, we believe of American origin, a number of pick plates of tempered steel are held in a hollow sliding block arranged with a vertical motion. A cap fits over the top of the pick plates, and is bolted to the hollow block; this cap is struck by the workman with a mallet, and the picks are forced into the stone. The bed of the apparatus is placed on the face of a mill-stone, and the hollow block is arranged to slide transversely in guides as well as vertically.

For quarrying purposes rock drills are employed largely; these are worked by steam or compressed air, and effect an immense saving, especially with very hard stone or granite.

For drilling a series of holes in a straight line by hand or power, the drills are usually raised and dropped by a series of cams, mounted spirally on a horizontal shaft. For cutting out holes in quarried stones, hollow cylinders of sheet iron are often employed; these are fitted to a cast-iron disc or tool-holder, which is screwed to the bottom of a drill spindle; this can be raised and lowered or advanced and retired at will, and according to the nature of the stone being worked. Notches are made in the lower edge of the cutter to admit the sand and water necessary for cutting (see Fig. 38).

For trueing the face and furrows of mill-stones, &c., and cutting down irregularities of surface, rubbing stones are employed. These are made of a composition in which carbonate, bort or diamond dust are used; the composition is usually moulded in the shape of a brick, which is fitted into a frame and rubbed over the surface of the stone to be cut by hand.

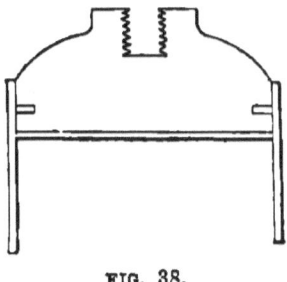

FIG. 38.

Amongst miscellaneous stone-working machines we must not forget to mention a very useful little apparatus for dressing up grindstones (Brunton's patent). This consists briefly of a slide-rest, which is bolted to the grindstone trough in such a position that the centre line of the tool spindle is in line with the centre of the grindstone spindle. The cutting is performed by means of a revolving disc of chilled cast iron or steel, as already described. In this case the cutting tool is set in motion by the revolution of the grindstone itself, against which it bears.

CHAPTER XVII.

CUTTING TOOLS.

Much has been written with reference to the forging and tempering of tools for cutting iron, wood, and other materials; but, so far as we are aware, nothing as regards tools for working stone. We need hardly say that, in the first instance, the steel employed should be of the highest possible quality, combining in its nature, as far as possible, toughness with hardness. Where sheet steel is used for scraping cutters, as in moulding machines, we prefer to use wrought-iron cutters, faced with steel, instead of solid steel, as they are easier to make and less liable to fracture, either when being hardened or when in work.

If chilled cast iron is used it should be of good quality, and not too short in the grain. In forging or hardening tools, it is important that they should be heated as evenly as possible. If one part of the tool is made thinner than the other, care must be taken that the thin part does not heat more rapidly than the rest, or it may become "burnt," or break off at the nose or cutting edge whilst in work. In heating the tools for tempering, they should be repeatedly turned over in the fire, and withdrawn from it now and then. In the case of moulding cutters, if the cutting edge is heating too rapidly, it should be pushed through the fire into cooler coals. The cutting or scraping edge should be well supported at the back by iron, which should not be ground to an acute angle, as is the custom with moulding and planing irons for working wood. If a number of tools or cutters are employed on the same machine, great care must be exercised in tempering as nearly alike as possible. This is a matter of great difficulty, and, as we have elsewhere pointed out, is the chief drawback to the use of this class of machines. Owing to the varied nature of the material operated on, no arbitrary rules as to tempering the tools of stone-working machines can be laid down. This can only be determined after trying both the stone and the steel, and must, therefore, be left to the experience of the workman. Speaking generally, for scraping and thin steel cutters, a light straw colour is usually suitable for stones of a moderate degree of hardness. Where the tools are stouter and the cutting edge more obtuse, the temper should be slightly harder in proportion.

It is important, in tempering any kind of tools or cutters, that there is a gradual shading of colour in the temper. If there is a distinct line between two colours towards the edge of the cutter, it will probably chip or break at this line. The point to aim at is to have the edge of the cutter tolerably hard, and this hardness to be gradually reduced the farther you go from the cutting edge, and the softer metal at the back will be found to strengthen and support it. In working stone, tools of extreme hardness are not always necessary or advisable, as with some kinds of stone the cutting edges will readily break or crumble away.

In machines where solid bars of steel are used, a temper colour of a deep blue, slightly tinged with violet, will with most stones be found suitable; the nature of the steel used must, however, be borne in mind, as some steels require greater or less hardening than others. The degrees of temperature necessary to effect the practical colour on hardened steel for the various purposes to which edge tools are applied are given by Templeton as follows : Chipping chisels, planing irons, hatchets, &c., and other percussive tools, 500° to 520°, light straw colour, a brown yellow or yellow slightly tinged with purple ; do. 530°, light purple ; springs 550°, dark purple, do. 570°, dark blue. As the colours appear and change colour slowly, ample time is afforded to see when the exact shade of colour has arrived, when the tool should be at once dipped and withdrawn several times— should boiled water be used for hardening—as this has a greater tendency to toughen the steel, than if it is plunged into the water and allowed to remain there till quite cold. We can strongly recommend the following method for hardening cast-steel tools : Take four parts of powdered yellow resin and two parts of train oil and carefully mix them, and add one part of heated tallow. The object to be hardened is dipped into this mixture red hot, and is allowed to remain in it until it is quite cold. Without having previously cleaned it, the steel is again put into the fire and is then cooled in boiled water in the ordinary way. After a little experience the eye of the workman will readily detect the exact shade of colour best suited to the steel and the stone he is working, and should he try the above process he will find the edges of tools thus hardened wear excellently. With some grades of steel salt water has been found to make an excellent hardening mixture.

It must be admitted there is a considerable amount of art in tempering steel tools properly, as owing to the varying amount of carbon contained in different samples of steel, the amount of tempering required varies accordingly, and the exact temper necessary can only be ascertained by one or more trials. It should be borne in mind that steel known as high-tempered steel is that containing a large amount of carbon, and low-tempered that containing little ; but as there are numerous gradations between high and low tempered, so will the points at which tempering should cease vary accordingly ; the folly, therefore, of treating all kinds of steel alike, which is sometimes done by the workman, will be at once apparent.

It may be taken as a rule that if it is necessary to heat steel so hot that when it is annealed it appears coarser in the grain than the bar from which it was cut, it may safely be concluded that the steel is of too low a temper for the required work, and a steel of a higher temper should be selected. A steel tool, when properly tempered and suited to the work in hand, should always be of a finer grain than the bar from which it was cut. We need hardly say that the action of tempering should be gradual, as the steel becomes toughened and less liable to fracture by slow heating and gradual softening, than if the process is performed abruptly. The steel being, of course, in the first instance, hardened, or made of a considerably higher temper than is necessary for the work it has to do, when the proper heat has been reached, the tool should be removed from the fire, and not allowed to "soak" with the blast off, as is sometimes done.

If the steel should crack with a moderate degree of heat, it probably contains too much carbon, or is of too high a temper for the purpose required. Try, therefore, a lower steel—*i.e.*, one containing less carbon, which will probably be found to be what is required. In ordering steel from the maker, always say exactly for what purpose it is required, as he will then be able to send a steel which he considers best suited to the requirements of each individual case. In many cases steel has been rejected as bad, when it has been of first-rate quality, but its temper, or the amount of carbon it contained, rendered it unsuitable to the work in hand. It may be taken as a general rule that steel of a high temper should be annealed at a low heat, in some cases even dark red will be suitable. We need hardly say nothing is so dear as "cheap" steel for tools, as, after putting a considerable amount of labour into it, it is soon thrown away by the steel rapidly wearing away, especially if it is operating on difficult materials like some kinds of stone. For welding cast steel, the following composition is effective: Boil together borax 16 parts, sal-ammoniac 1 part, over a slow fire; when cold, grind into a powder; and use the same as sand. Bear in mind in forging, welding, or tempering steel tools, that an excess of heat over what is absolutely necessary is detrimental, as it opens and makes the grain of the steel coarse; also that the steel should be heated as regularly as possible, as irregular heating causes fractures, and irregular grain and strength. If a tough temper is required, the cooling or letting down should be as slow as possible.

In tempering tools, where many are employed, a frame with holes in it to equalise the distance tempered may be used. This is fitted in the hardening tank, and allows the tools to drop any desired distance into the water or tempering mixture.

As regards the shape of the tools for dressing stone, we have already spoken; but if a machine similar to an ordinary planing machine for iron be employed, in which the tool-box is stationary, and a point tool used, it will be found well—at any rate with stone of a moderate degree of hardness, and where tolerable finish is required —to make the cutting point of the tool as nearly on the moving plane of the work as possible. The best cutting angle for all kinds of stone will generally be found between 45° and 90°. The heel of the tool should only be raised slightly above the front or cutting edge. As much metal as possible should be left to support the cutting point, allowing only sufficient clearance to give a proper angle to the cutting edge; it will be found necessary to vary this angle somewhat, according to the nature of the stone being worked. If the tool, when in work, is found to spring much, or dig into the work, it will require, in the first place, strengthening, and the cutting angle will require modification; at the same time, if too great a cut is taken, the tool will have a tendency to dig into the stone. It will—as with some tools for cutting iron— with difficult stones, be in some cases advisable to make the tool so that the cutting edge is behind the fulcrum or point of support, and should any very hard spots be found in the stone, such as crystalline or fossil deposits, it will, as a rule, yield to them instead of breaking. With stones easy to work, double-nosed tools may be used, the tool box and speed of the travelling table carrying the stone being so arranged that a cut may be taken during both the back and forward traverse of the table. The width of the cutting tool may be varied according to the nature of the stone. The easier the stone, the wider the tool; but for a roughing cut, where much stone has to be removed, it should not exceed about $\frac{5}{8}$in.

In those stone-dressing machines in which the cutters are reversed so that each edge of the tool is brought into contact with the stone, comparatively little sharpening is required, as the rubbing action of the stone itself is found to keep the edge of the tool in tolerable condition. In machines in which the tools are fixtures, or are non-reversible, the condition of their cutting edges should be carefully attended to. In stone-converting works of large size it will be found well to appoint an intelligent man to sharpen and look after the whole of the tools, and if he can forge and temper them as well, so much the better, as he will by experience be able to secure what is a matter of great importance—*i.e.*, a tolerably uniform temper, and one suited to the stone being worked. For grinding, an ordinary Newcastle grindstone with Water-of-Ayr stone attached, will be found suitable. It will be better to sharpen moulding irons by grinding than by filing, as the process of softening the steel and re-hardening is at the best uncertain, and the quality of the steel is certainly not improved.* For grinding moulding irons some half-dozen Bilston grindstones about 14in. diam., and of thicknesses varying from 1in. to $2\frac{1}{2}$in., are generally employed. These are mounted on a spindle fitted in a cast-iron trough supplied with water. The stones are turned up to suit the shapes of the moulding irons most commonly in use.

For grinding down and shaping moulding irons the emery wheel will be found much more rapid and economical than either files or grindstone; prejudice, and the introduction of emery discs of an inferior quality have, however, in a measure retarded their general use. If an emery disc of a good quality is used, and a fair trial given to it, it will, we feel sure, be found a most labour-saving and valuable implement. Emery wheels should not be used for putting a finishing edge on irons, but for reducing the back of the cutting edge and the iron generally to the desired profile. If much steel has to be ground away it

---

* See "Saw Mills: Their Arrangement and Management," by M. Powis Bale.

will be found better to go over the iron two or three times, than to put too great a pressure on the wheel, as the iron will become of a dark blue colour, and hard and liable to crumble away at the edge when in work. In working emery wheels, for whatever purpose they are employed, great care should be taken that they are run at a correct speed; for general purposes a speed at the periphery of 4,500ft. to 6,000ft. per minute, but not above, will be found suitable. If emery wheels are run too fast they may fly to pieces, and if too slow they will not cut properly. In selecting a steel for stone-working tools do not choose one too short in the grain; as we have before remarked, one combining a certain amount of toughness with hardness will, as a rule, be found most suitable. Some steels will stand almost a white heat and yet temper well, whilst others have to be worked at a low red heat, so no arbitrary rules as to tempering colours can be made; it simply resolves itself into a matter of actual practical experience. When a steel suitable for working the stone in hand has been found, stick to it, and do not keep constantly changing.

For planing stone by fixed tools, either solid bars of steel are forged to the desired cutting shape, hardened and tempered, and mounted in a tool box, or, more recently, in lieu of bars, small pieces of steel mounted in tool holders, have been employed with advantage. The tool holders are of wrought iron, and the cutting tool is held in position by means of set screws, wedges, dovetail grooves, or other similar plans. In some planing machines steel of tapered section is employed, and this can be used without much forging; the desired cutting angle for stone being obtained by grinding. Bars of steel of a trapezoidal section—that is, the one edge of the bar is somewhat wider than the other edge—are also used for tools for moulding and planing usually in machines arranged with a vertical cutter barrel. One end of the steel bar is ground obliquely, so that the bar at its wider edge stands somewhat further forward than at its back edge, the front wide edge forming the cutting edge of the tool. With some kinds of stone the end of the tool is ground square, and set a little obliquely in its rest. As we have elsewhere remarked, we prefer a system of rolling cutters for dressing stone, to a plane surface; we shall not notice at length, therefore, this form of machine tools, which are nearly allied to those used for iron planing. Whatever tools are employed, care should be taken that they are of sufficient strength and sufficiently self-supported, that they do not spring whilst working.

When, by experiment, the best form for the cutting angle and clearance angles of tools for dressing various stones has been ascertained, it is of great importance that a registry of them be kept, and care taken that they are exactly adhered to in all tools used for that class of stone. This can be secured by adopting a mechanical system of re-grinding, and by making sets of sheet-steel standard angle gauges, to which the various tools can be ground. The rule of thumb should in all cases be avoided, as no matter how carefully carried out, where several cutters are employed a certain amount of irregularity must exist, the result being the stone requires an extra amount of finishing on the rubbing-table, when rubbed faces are required. A graduated adjustable slide-rest should in all cases be fitted to the grindstone, as it is impossible to secure accuracy by means of hand-grinding, and as we have elsewhere remarked, it will pay well to have one man especially to keep the cutting tools in order, and from experience he will soon learn exactly what is required, and be able to judge from the nature of the stone what width of cut may be safely taken. Where possible with easy working stones, we recommend broad cuts, at any rate for finishing purposes, say up to $1\frac{1}{2}$ wide, as a more accurate surface is obtained, and a considerable saving in time effected.

If ordinary straight tools are employed for planing or turning they will occasionally be found to spring and dig into the work, especially if a deep cut be taken; this can generally be remedied by either cranking the cutting end of the tool or increasing its section; if the tool is strong this cranking may be done by grinding a hollow in the top side of the bar near the end. Thus, when in work, the tool having a certain amount of pliability, if it springs at all it will spring outwards, and not dig into and score the work. It is very important that the cutting angle of the tool be suited to the stone being turned, and when this is found it will in the end be the most economical to use steel of a section sufficiently strong to prevent its springing appreciably in either direction. If a short cutting tool and a tool holder be employed, the former should be strongly supported.

One of the best and simplest forms of tool holders for turning lathes with which we are acquainted is that of

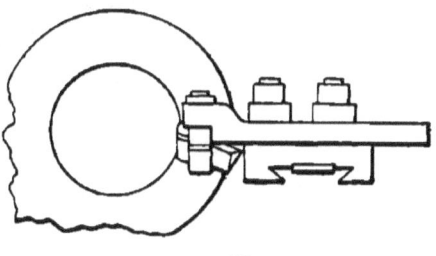

FIG. 39.

Messrs. Smith and Coventry, which we illustrate by Fig. 39. As will be seen from the sketch, the tool holder is of square section with a boss forged on its front end. A bolt passes through this boss; through the centre of the bolt immediately above its head, which is placed below the boss, is cut a vertical rectangular slot, and through this is passed a piece of steel of taper section, which performs the cutting. The tool is held in position by being fitted into slots cut into the upper face of the head of the bolt, and in a thick washer placed above the cutting tool and immediately below the boss of the tool holder. These slots are cut so that the steel tool stands with its upper face inclined to the horizontal, and gives the usual amount of top rake required. The nose of the tool can be ground to any desired shape, and the tool can be swivelled to any horizontal angle with the axis of the work, whilst the tool holder itself always remains perpendicular to that axis. One of the advantages claimed for this tool holder is that the cutting tool can be instantly set to any desired horizontal inclination, according to the shape of the work being turned. The shape of the cutting tool allows it to be ground to any desired rake, and gives it stiffness in working. With reference to the circular cutters already noticed, for hard limestone steel is necessary, but for gritstones, sandstones, and freestones, chilled cast-iron answers well, as it is as hard as steel and very inexpensive. The steel cutters are made in the form of flat discs or cones, but the cast-iron cutters always in the form of cones.

In practice, where a tolerable thickness of stone has to be removed, say, about 1in. in depth, it will be found preferable to take two cuts instead of one, as with some classes of soft stone the cutting tool will be found to dig downwards and "pluck" the stone. This, however, is more particularly the case with straight steel tools; at the same time the friction on the tool is largely increased.

CHAPTER XVIII.

NOTES ON MANAGEMENT OF A STONE WORKS.

In the management of the yard or mill for stone conversion, it is important that ample appliances for lifting and transporting the stone are to hand. In a large works a steam overhead travelling crane can be used with advantage for lifting up to, say, 20 tons. The frames and crossbeams may be made of oak or pitch pine, but for heavier weights iron girders should be used. A very steady traverse of the crane may be obtained by using worm and worm-wheel gearing from the engine crankshaft, working spur and bevel gear fitted to the wheel of the travelling carriage. The hoisting gear should be arranged with single and double purchase. The different transverse motions should be arranged to work either separately or simultaneously. A capstan or warping drum can be fitted with advantage to the framework of the crane, and utilised for bringing the stone trucks into position for unloading. This capstan should be arranged to drive independently of the travelling wheels.

In establishments of moderate dimensions a handpower overhead travelling crane may be sufficient. It will be found most convenient to work the lifting, transverse, and longitudinal travelling motions from below. To prevent accidents a powerful brake and catch arrangement should, in all cases, be fitted. The works should be amply provided with crab winches fitted with brake, pulley blocks, stone bogies, stone shears, chain slings, louises, and several sets of masons' and quarrying tools. In lifting from quarries of considerable depth, perhaps the best plan is to use steel wire rope arranged to pass from the crane to a winding barrel. The necessary gearing can be fitted to a crane in any suitable postion. This plan will be found simple and expeditious, or strong timber stages fitted with travelling frames may be used; in this case upon the travelling frames powerful traversing crabs are employed. These should, in all cases, be fitted with double purchase gearing and powerful brakes, and best tested lifting chains. The spindles should be bushed with gun-metal or phosphor bronze.

In establishments where many moulded or plain bevelled surfaces are produced, it is important that the travelling table carrying the stone, or a cradle mounted on it, are capable of adjustment laterally to various degrees of inclination, so that the stone can have its edge or face inclined, according to the angle or pattern of the moulding to be cut on it.

In a stone-converting, as in all other works, manual labour should be reduced to the lowest limits by the introduction of all kinds of labour-saving appliances, and by having the building or yard, and its arrangements, well suited to the work to be performed. The division-of-labour system should, as far as possible, be carried out; when a steady and uniform business is in vogue, this is particularly to be recommended, as it will be found that the workman, by constant repetition of the same work, will increase the output of his machine 15 or 20 per cent.*
As regards the labour employed, the highly-skilled and,

---

* See "Saw Mills: Their Arrangement and Management," by M. Powis Bale.

consequently, the highly-paid workman is, as a rule, the cheapest; the first difference in cost being soon counterbalanced by an increased output from the machine and a better average quality. We are strongly in favour of piecework where it can be managed; it also has the additional advantage of offering a premium to the operator for keeping his machine and tools in constant use and in the highest state of efficiency. Accuracy of workmanship should in all cases be insisted on. If a number of dressing tools are employed on one machine, care should be taken that they are all ground to the same angle, and tempered as nearly alike as possible.

We need hardly say it should be the aim of the management to waste as little stone as possible; all blocks should therefore be carefully inspected and measured, so as to work them up to the best advantage. The cost price of work should be constantly taken out, and it is advisable that the workmen should in all cases book their time on all the work they are employed on without exception; although in some cases this may not be altogether necessary, it acts as a check on day workmen if they understand their time is likely to be dissected. All stores should be charged against each machine, and it will be found well to give each a number.

It is important that a thoroughly good works foreman be employed—a man who knows what to do and how to do it—and we are in favour of allowing a foreman, in addition to his weekly wages, a small bonus or commission on results above a certain point.

Owing to the dust floating about a stone-works, especial care must be taken that the various bearings are well lubricated and protected. As regards a lubricant, we can from experience recommend either of the following mixtures: Good lard oil, 75 parts; plumbago, or sulphur powdered very fine, 25 parts. If the spindles are light, the amount of plumbago may be reduced; should they be heavy, running at a low speed, and subject to constant and great pressure or strain, the plumbago may be increased with advantage up to 40 parts. Oils containing acids or alkalies and low grades of mineral oils, should in all cases be avoided, as they are acrid in their nature, and their oleaginous properties are small. Amongst the fatty lubricants, sperm oil, lard oil, and Russian tallow, hold the foremost rank, and if they are carefully used they will last much longer than oils of low grades, and the bearings also will keep in better condition, as thin poor oils are rapidly absorbed or expelled.

The shafting employed should be carefully arranged and fitted, and short centres should be avoided. Where large power is transmitted from a pulley, a bearing should be placed on either side of it. The plummer blocks used should be adjustable; the best with which we are acquainted is made on the principle of the universal or ball and socket joint. The advantage of this plan is that, in whatever direction the shaft may incline, there is an equal wear or strain upon the whole surface of the bearing, and should the plummer blocks be set somewhat out of truth, the ball and socket joint allows the bearings to adjust themselves in line. As we have before remarked, we prefer the shafting to be mounted on standards and fixed underground, as it is easily got at and is out of the way.

All driving belts should be kept pliable. Tight belts should be avoided. If more driving power is required, increase the width of the belt. An application of Tanner's dubbin for leather and linseed oil varnish for cotton driving belts will make them pliable and increase their driving power. Belts should be kept free from moisture.

Wherever possible, the machinery should be so planned that the belts never have to run in a vertical line. Avoid the use of resin and other mixtures sold to increase the grip of the belt, as in most, if not all, cases they act injuriously on the leather.

A mixture of mutton fat and beeswax will be found a capital dressing, and will not injure the leather. Castor oil is also an excellent dressing for belts; it should be mixed in equal proportions with tallow or oil.

Profile templates of the various stone mouldings generally in use should be made.

In managing a stone or any other works economically it is important, if steam power is used, that the steam is not wasted. This can, in a measure, be guarded against by the employment of an engine fitted with an automatic expansion slide, which regulates the consumption of the steam to the power required to drive the machines actually in motion. The draught to the boiler should be regulated by a damper; this, if possible, should be arranged to work by steam automatically, as it requires no attention, and is regular in its action, and effects a considerable saving over the old form of slide-damper worked by hand, as its regular working is often neglected by the fireman in charge. A steam-damper can be arranged to act at any desired pressure of steam; and, as the fire is automatically damped when that pressure is reached, a very considerable saving in fuel is thus effected. A feed-water heater should also be employed—if in combination with a filter so much the better. If the water contains iron or acid sulphates it must be purified and softened before use.

As we have elsewhere remarked, it is of much importance to the health of the workman that the stone dust is kept down as much as possible, by being damped and cleared away at stated intervals, which can be judged from the nature of the stone, some making much more dust than others. As illustrating the injurious effects of stone dust, Wiseman, in his book on Surgery, relates a case in which a stone-cutter's man had the vesiculæ of his lungs so stuffed with dust that, in cutting, the knife went as if through a heap of sand.

It is important that where stone has to resist the effects of the atmosphere, or to be polished, that its surface should be dressed true and be "unstunned." If the machine therefore that is employed to dress it, leaves the stone with an irregular "plucked" face it may be concluded, that either the type of the machine employed is unsuited to the stone or the stone is not suitable for machine conversion. As regards the best types of machines, speaking generally, for dressing and planing, especially for the harder classes of stone, we are in favour of the use of revolving chucks and cutters, and for moulding purposes, a combination of revolving roughing and stationary scraping cutters.

CHAPTER XIX.

STONE-WORKING—HAND LABOUR VERSUS MACHINERY.

ALTHOUGH there is little doubt that machinery has a vast economical advantage over hand labour in the conversion of stone, it is—owing to the varying nature of the work and prices paid—a somewhat difficult task to compare the two systems by absolute figures. In calculating and comparing the cost of machine and hand labour on the one hand, we must take the cost of steam power, labour, renewal of cutters, oil, &c., and interest on capital sunk, with a certain allowance for depreciation. On the hand-labour side, we must put cost of labour, tools, &c., not forgetting the increased length of time necessary to produce a given amount of work, which means an increased rental.

Taking, first, ordinary stone dressing : the number of superficial feet which can be dressed by a machine in a day will, of course, depend greatly on the nature of the stone being worked, and in the way the machine is kept constantly supplied with stone. It is important, therefore, in order to secure the best results, that little or no time is lost in fixing and unfixing the stone. The size of stone being dressed will also affect the number of feet produced in a day, as with a machine it will take almost as much time to dress a small stone as a large one.

With the ordinary stones used in building construction of a moderate degree of hardness, a well-designed machine should, on a fair average, dress about 80 superficial square feet per hour; the cost of this, allowing two men and one boy to supply stone and attend to the machine, would amount to about 2s.; whilst an average price to dress the same by hand would be about 5s., leaving a large profit and surplus for contingencies.

In the case of hard grit stones, an output of rather more than one-half the above—say, 220 superficial feet per day of 10 hours—should be considered an average day's work.

The makers of Brealey's Patent Stone Dressing Machine assert that one of these machines, to take in 4ft. wide by 2ft. high and 9ft. long, will tool or face from 300 to 500 superficial feet of flags or landing, according to quality, per day of nine hours, at a total cost of 12s. per day. If we take the medium of, say, 400ft. per day, which if worked by hand labour would cost $1\frac{1}{2}d.$ per foot, we find that the cost would be £2 10s.; whereas by the aid of a machine the work can be better done for about one-fourth the sum.

Messrs. Brunton and Trier state that their machine has dressed to a perfectly good surface in sand and gritstone 1,296 superficial feet in 49 hours, and in granite as much as $17\frac{1}{2}$ square feet in an hour.

A comparison of the cost of working moulding in stone by hand and by machinery has been published by the *Builders' Trade Circular*. The work on which the cost price was taken was a number of staircase steps in white Portland stone, 5ft. long by 7in. deep, with a moulding girting about 12in. They were cut one out of the other by steam stone saws, the cost of sawing being less than 2d. per foot. The moulding machine used was by Rotheroe, and would take a stone 10ft. long by 5ft. 6in. wide. It was a modification of an ordinary planing machine for working iron, and was fitted with a travelling bed with reversing action, and with two double tool boxes, so as to operate on the stone, if required, during both the backward and forward traverse of the table. The steps were reduced on the machine, with an ordinary point tool, to within a quarter of an inch of the correct size, and then scraped or finished with a tool made of sheet steel shaped to the exact profile of the moulding required. The number of steps thus moulded by the machine during several months averaged 18 per. diem, giving as the day's work about 90 superficial feet of moulding. Two men worked the machine, and the steel tools were made on the premises, and the cost of these was about £2 5s. per week; one smith being pretty constantly, though not entirely, occupied in making, altering, repairing, sharpening, steeling, &c., as required. The first cost of the machine was £300, exclusive of foundation and labour fixing.

From these data, the cost of a day's work was given; but several of the items would vary in different localities, such as the cost of labour and steam power. The estimate of cost was as follows:—

	£	s.	d.
Two labourers, one day each, 20 hours at 5½d. per hour. (N.B.—They were active, intelligent men)	0	9	2
Tools as above, per week 45s., per day	0	7	6
Steam power, say about	0	5	0
Interest upon outlay at 5 per cent., say	0	3	6
	£1	5	2

At the ordinary price of London labour, 90ft. of Portland moulding, at 1s. 2d. per foot prime cost for hand labour, would amount to £5 5s. 4d. Assuming the machine does this work at £1 5s. 2d., a very large margin for profit and contingencies is left. By the ordinary system of London measurement, all carving and setting is put on the price of the stone; the price, 1s. 2d. per foot, as quoted above, being for the labour of moulding only: the comparison, therefore, is absolute. Whatever the nature of the work, nearly the same superficies of moulding was worked, and the day's work given is the average of many days.

In different localities, of course the conditions will vary. We think, however, the cost of machine moulding, as given above, is somewhat low, especially the wages of the attendants: and in practice we think it will be found more economical to employ one man, of a higher class than a labourer, who can sharpen and adjust his cutters, and be able to overcome any little difficulties that may arise owing to the varying nature of the stone worked, &c. However, should the cost of producing the 90ft. of Portland moulding amount to 30s.—which sum, we think, will be found nearer the mark, except under special circumstances—a very large margin of profit is still left.

Another advantage that machine moulding will be found to have over hand moulding is that the work generally is better done, the lines being perfectly straight, and the edges beautifully sharp; and should a large number of pieces of one moulding have to be worked they will be found absolutely alike in section.

The amount of work turned out per day will of course vary with the size and nature of the stone, section of moulding, and if the supply of stone is constant, and also if many changes of work are necessary. For working smaller stones, at any rate—such as stair steps, stringcourses, sills, &c., and builders' work generally—a moulding machine with a vertical cutter barrel is undoubtedly to be preferred, as it is more easily worked, and the tools more readily adjusted or exchanged. The first cost, also, should be less than a machine with a horizontal barrel.

An advanced type of vertical barrel machine should be capable of cutting a straight moulding girting, say 15in., on a piece of Portland 4ft. long, similar to sketch, Fig. 40, in thirty minutes: this we ourselves have done.

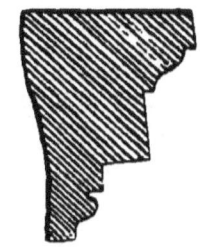

FIG. 40.

The following statement of work done by a Duplex barrel moulding and planing machine working in Portland stone was taken from the book of a workman:—string 1ft. deep, 7in. projection, 50ft. run per day of 12 hours. Moulded steps, 24 per day worked both sides, with 4 steps on the machine at once.

Window jambs 6in. face, full of mouldings, worked both sides, 150ft. run per day.

As showing the large amount of work that may be produced by the use of tools with a rolling contact, in preference to those with a dead contact, we have, on the authority of Messrs. G. and J. Fenning, of the Shap Granite Works, Westmoreland, that they turned in a lathe, with Brunton's patent cutters, 42 granite columns, of all sizes above 8in. diameter, amounting to 1,100 superficial, in 383 hours; and they estimate one mason would have spent 4,428 hours in doing the same work. In the case of columns under 8in. diameter, the lathe turned 71 lineal feet in 114 hours; to do the same amount of work, a mason would have been occupied about 648 hours.

Although a very large saving can be effected over hand

labour in the conversion of stone by the employment of machinery, *it is important that the machinery selected or tools used should be suited to the nature of the stone and the work to be carried out.* If in every case, before purchasing, the builder or quarry owner had the stone he chiefly used or sold tried on several types of machines, he would then be able to select the one best suited to his requirements; and, in some cases, much loss and annoyance would be avoided.

In comparing the cost of horizontal-blade machine and hand sawing, the saving effected—although considerably in favour of machine conversion—will not be found so large as in moulding or dressing machines. The number of superficial feet sawn will, of course, depend largely on the nature and size of the stone and the number of saws employed. In sawing soft stone, such as Bath or Portland, the cost of machine sawing may be fairly set down at about $2d.$ per foot, and the output with 6 saws at about 72 square feet per day of 10 hours. The cost of hand sawing will generally average about $3\frac{1}{2}d.$ per foot; but the work is not so truly done, and therefore requires more finishing on the rubbing table. At the same time the process is much slower. For sawing soft stone, steel-toothed saws have latterly been introduced with very satisfactory results as regards production. The harder kind of freestone will not average more than about one-half the above output.

Circular saws will be found much more expeditious for sawing stone than straight ones. They, however, have the disadvantage of taking greater power to drive, and at the same time the waste of stone is greater. Notwithstanding these drawbacks, the author can strongly recommend circular saws for converting Portland, Bath, York, and similar classes of stone. A well-constructed machine of this class will cut from 150 to 250 running feet per day of 10 hours; the output, of course, varying according to the nature of the stone being operated on. Should circular saws be employed for dressing the faces of the stone, they can be made to turn out from 3in. to 9in. run per minute, and for this purpose they are really very valuable. The cost of rubbing or facing building stones on a large well-constructed machine may be set down at from 1d. to 3d. per superficial foot, according to the nature of stone. In turning granite with revolving cutters, a well-constructed lathe will give an output of at least 3 times the amount that can be produced in a lathe in which the turning tool is a fixture; in soft stones, however, the difference in output is not so great.

The Author desires it to be understood that the figures as regards the outputs of various machines were obtained in most cases from the makers, and he declines therefore the responsibility as to their correctness, with the exception of the Hunter Duplex Moulding and Planing Machine and Circular Saws, the output of which he has taken from absolute work done.

CHAPTER XX.

NOTES ON MASONRY, RECIPES, ETC.

In districts where stone is abundant and is largely used for building, ordinary walls are built of rubble stone, and measured by the cubic yard, or some other local standard, in Ireland by the running perch (21 ft.) of a given height and thickness, or by the square perch of 21 ft. super., at a standard thickness of 18 inches.

A cubic yard of rubble masonry will, as a rule, require $\frac{1}{5}$ cubic yard of mortar and $1\frac{1}{5}$ cubic yard of stone.

In the West of England, the square perch is employed, of 18 ft. super.; at a standard thickness of 2 ft., to which walls of any thickness are reduced. In other parts masonry walling is measured by the rood of 36 square yards, or 24 cubic yards, the standard thickness being 2 feet, or by the rod of 272 feet super., as in brick wall at a standard thickness of 18 in. or 2 feet.

The superficial content of the surface work is paid for separately by the square foot, yard, or rood of 36 square yards, including jointing or pointing, and such squaring to beds and joints as may be required, quoins, &c.; of selected stones being often paid for at so much extra per foot run.

Block stone, for the cut stonework in buildings, is generally sold at the quarries, or delivered by the cubic foot, or in large rough blocks by the ton, of from 13 to 17 cubic feet, 1 in. being allowed each way for irregularities and waste.

Quantity of stone of different kinds, equivalent to a ton in weight.

	FEET, CUBE.
Vein Marble	. 13
Statuary Marble	. . 13¼
Granite	. 13½
Purbeck	. . 14
Yorkshire	. 13 to 14½
Craigleith	. . 14¾
Portland	. 15 to 16
Derby	. . 15
Bath	. 16 to 17

INCHES.	FEET, SUPER.
2½ York Paving	. . 70
3 York „	. 58
2½ Purbeck „	68
3 Purbeck „	. 56
3 Granite „	54
6 Granite „	. 27
7 Granite Curb	. 23

The following constants, chiefly for work on stone, are taken from the "Student's Practical Guide to Measuring," by E. Dobson. The factor to be applied is the daily rate of wages for a mason.

Labour on Portland or similar stone, per foot superficial.

N.B.—Sawing to be taken as half plain work.

	per foot super. DAY.
Plain work to bond stones	·14
„ to beds and joints	. . ·181
„ rubbed face	. ·209
„ „ circular	. ·291
Sunk work, rubbed	·25
„ „ circular	·313

M 2

	DAY.
Moulded work, rubbed	·292
" circular	·417
Circular work to shafts of columns having the neck moulding or part of the base worked in the same stone	·334
Circular—Circular or spherical work to domes or balls	·5
If rubbed, add extra	·049
Labour, squaring and laying York or Purbeck paving	·021
If in courses, add	·01
Taking up, squaring, and relaying old paving	·042

PAVING.

One ton of 6-inch granite paving will cover 4 yards super.
"   7-inch   "   " $3\frac{1}{2}$ "
"   8-inch   "   "   3   "
"   9-inch   "   " $3\frac{1}{3}$ "
One ton of pebble paving will cover from 4 to $4\frac{1}{2}$ yards.
"   ragstone   "   "   "   5 to $5\frac{1}{2}$ "

KEYSTONES.

To compute the depth of keystones for segmental arches of stone (Trautwine).

First class of arch.—·36 $\sqrt{}$ of the radius at the crown.

Second class of arch.—·4 $\sqrt{}$ of the radius at the crown.

Brick or rubble.—·45 $\sqrt{}$ of the radius at the crown.

In viaducts of several arches. Increase the above units to ·42, ·46, and ·51.

MARBLE, WHITE.—Specific gravity, 2,706; weight of a cubic foot, 169 lbs.; weight of a bar 1ft. long and 1in. square, 1·17 lb.; cohesive force of a square inch, 1811 lbs.; extensibility, $\frac{1}{1934}$ of its length; weight of modulus of elasticity for a base of an inch square, 2,520,000 lbs.; height of modulus of elasticity, 2,150,000 ft.; modulus of resilience at the point of fracture, 1·3; specific resilience at the point of fracture, 0·48 (Tredgold); is crushed by a force of 6060 lbs. upon a square inch (Rennie).

STONE, PORTLAND.—Specific gravity, 2·113; weight of a cubic foot, 132 lbs.; weight of a prism 1 inch square and 1 foot long, 0·92 lb.; absorbs $\frac{1}{16}$ of its weight of water (R. Tredgold); is crushed by a force of 3729 lbs. upon a square inch (Rennie); cohesive force of a square inch, 857 lbs.; extends before fracture $\frac{1}{1789}$ of its length; modulus of elasticity for a base of an inch square, 1,533,000 lbs; height of modulus of elasticity, 1,672,000ft.; modulus of resilience at the point of fracture, 0·5; specific resilience at the point of fracture, 0·23 (Tredgold).

STONE, BATH.—Specific gravity, 1·975; weight of a cubic foot, 123·4 lbs.; absorbs $\frac{1}{13}$ of its weight of water (R. Tredgold); cohesive force of a square inch, 478 lbs. (Tredgold).

STONE, CRAIGLEITH.—Specific gravity, 2·362; weight of a cubic foot, 147·6 lbs.; absorbs $\frac{1}{63}$ of its weight of water; cohesive force of a square inch, 772 lbs. (Tredgold); is crushed by a force of 5490 lbs. upon a square inch (Rennie).

STONE, DUNDEE.—Specific gravity, 2.621; weight of a cubic foot, 163·8 lbs.; absorbs $\frac{1}{511}$ part of its weight of water; cohesive force of a square inch, 2661 lbs. (Tredgold); is crushed by a force of 6630 lbs. upon a square inch (Rennie).

STONE-WORK.—Weight of a cubic foot of rubble-work, about 140 lbs.; of hewn stone, 160 lbs.

SLATE, WELSH.—Specific gravity, 2·752 (Kirwan); weight of a cubic foot, 172 lbs; weight of a bar 1 foot long and 1 inch square, 1·19 lbs.; cohesive force of a square inch, 11,500 lbs.; extension before fracture, $\frac{1}{1370}$; weight of modulus of elasticity for a base of an inch square,

15,800,000 lbs.; height of modulus of elasticity, 13,240,000 ft.; modulus of resilience, 8·4; specific resilience, 2 (Tredgold).

SLATE, WESTMORELAND.—Cohesive force of a square inch, 7870 lbs.; extension in length before fracture, $\frac{1}{1640}$; weight of modulus of elasticity for base 1 inch square, 12,900,000 lbs. (Tredgold).

SLATE, SCOTCH.—Cohesive force of a square inch, 9,600 lbs.; extension in length before fracture, $\frac{1}{1645}$; weight of modulus of elasticity for a base 1 inch square, 15,790,000 lbs. (Tredgold).

ALABASTER CEMENT.—Plaster of Paris, 1 part; yellow resin, 2 parts; mix and apply hot, warming the faces of the fracture or joint; or sulphur or shellac, melted with plaster of Paris, or plaster of Paris alone.

GRANITE CEMENT.—Gum dammar, marble dust, felspar; the mineral ingredients are reduced to an impalpable powder, and the mass is incorporated by gradual heating. It should be applied warm to the warmed faces of the fractured portions. The black felspar is preferably used, to prevent the detection of the joint. For a hard cement: dried and pulverized clay, 8 parts; clean iron filings, 4; peroxide of manganese, 2; sea-salt, 1; borax, 1; triturate, reduce to paste with water, use immediately, heat after using.

MARBLE CEMENT. — Plaster of Paris steeped in a saturated solution of alum, and recalcined. Mix with water, and apply as plaster of Paris; this cement is susceptible of a high polish, and may be coloured to imitate marbles. A composition of gum-lac, coloured to suit the occasion.

The rust cement is also used, composed of hydrochlorate of ammonia, 2 ; flour of sulphur, 1 ; iron filings, 16.

For coating insides of cisterns, pulverized baked bricks, 2 ; quicklime, 2 ; wood-ashes, 2 ; olive-oil to make a paste.

For stone seams and joints, pulverized tiles or hard brick, 6 ; white-lead, 1 ; litharge, 1 ; oil to compound.

Hydraulic cement, 12 ; triturated chalk, 6 ; fine sand, 6 ; infusorial earth, 1 ; all mixed with soluble soda-glass.

STONE CEMENT.—Fine sand, 20 parts ; litharge, 2 ; quicklime, 1 ; linseed oil to form a paste.

A mineral compound for uniting stone and resisting water is made by mixing 19 lbs. of sulphur, with 42 lbs. of powdered glass or stone ware ; over a gentle heat the sulphur melts, and the whole is stirred till a homogeneous mass is obtained, when it may be run into moulds. It melts at 248° Fahr., and becomes hard as stone, and will resist boiling at 230°. Fahr.

CEMENT FOR JOINING STONE.—Sulphur, bees-wax, and resin, equal parts ; the sulphur and resin are to be melted together : the wax is then added, and the whole intimately mixed. The edges of the stones to be joined are to be gently heated, and in that state anointed with mastic, and then pressed tightly together until they are quite cold.

ROMAN CEMENT. — Parker's analysis. One part of common clay to $2\frac{1}{2}$ parts of chalk, set very quick.

CONCRETE.—Eight parts of pebble, or pieces of brick, to 4 parts of scrap river-sand ; and 1 part of lime mixed with water, and grouted in, makes a good concrete.

LIME MORTAR.—One part of river-sand to 2 parts of powdered lime, mixed with fresh water.

HYDRAULIC MORTAR.— One part of pounded brick-powder to 2 parts of powdered lime mixed with fresh water. This mortar must be laid very thick between the bricks, and the latter well soaked in water before laid.

HYDRAULIC CONCRETE, by Treussart. Thirty parts of hydraulic lime, measured in bulk before slaked; 30 parts of sand; 20 parts of gravel; and 40 parts of broken stone—a hard limestone.

CEMENT FOR STONE AND BRICKWORK.—Two parts of ashes, 3 of clay, and 1 of sand, when mixed with oil, will resist the weather.

WATERPROOF FOR CISTERN STONES.—(1), whiting 100 parts; resin, 68; sulphur, $18\frac{1}{2}$; tar, 9; melt together. (2), sand, 100 parts; quicklime, 28; bone ashes, 14; mixed with water.

ARTIFICIAL STONE can be made of clean sand and silicate of lime.

PRESERVATION OF STONE. — Zyerelmey's plan is to saturate with a silicate, and apply asphaltum varnish. Ransome's is to saturate with silicate of soda, and then with chloride of calcium. Hibble's plan is to paint with a compound of ground lime, turpentine, flax seed oil, silicate of lead, and burnt copperas. Davies proposes sulphur and flax seed oil. Barff and Sullivan, treatment with alumina, carbonate of zinc, and silicate of potash. Bernay's, fluo-silicic acid, washed with alkaline solution. Rust and Mossop; solution of caustic barytes, washed with fluo-silicic acid. Gros's plan; a paint of wax, 10 parts; oil, 30; litharge, 1; heated to 212° Fahr. Spiller's;

superphosphate of lime, followed by ammonia (for Magnesian limestone). Crooke's ; fuller's earth in a dilute solution of hydrofluoric acid.

For tempering mill bills and other tools for working hard stone, we have heard the following plan very well spoken of. 1.—Take 2 gals. of rain water, 1 oz. of corrosive sublimate, 1 oz. of salammoniac, 1 oz. of saltpetre, $1\frac{1}{2}$ pints of rock salt. The picks should be heated to a cherry red, and cooled in the bath. The salt is intended to give hardness and the other ingredients toughness to the steel. 2.—After working the steel carefully prepare a bath of lead heated to the boiling point. In it place the end of the pick to the depth of $1\frac{1}{2}$ inch until heated to the temperature of the lead, then plunge immediately into clear cold water. The principal requisite in making mill picks, &c., are (1), good steel; (2), work it at a low heat; (3), heat for tempering without direct exposure to the fire, the lead bath is intended as a protection against the heat, which is usually too great to temper well.

APPENDIX A.

SINCE the publication of the first edition of this book, some progress has been made in stone-working by machinery, both by the introduction of new machines and by improvement of details and methods of working the older ones.

HORIZONTAL SAW FRAMES are now largely built of steel and iron instead of wood and run at a higher speed. Instead of single cranks or pendulums for driving, double cranks and connecting rods are largely employed, these are usually connected to the swing or saw frame either at the centre or back end, thus obtaining the advantages of long connecting rods with increased steadiness in working in a lessened ground space. Improved self-acting downward feeds, and new forms of sanding boxes for feeding the grit to the blades, have been also introduced. In addition to sharp sand various preparations of powdered steel and chilled iron, &c., are used for cutting; these are known under various titles such as "chilled shot," "diamond grit," "krushite," &c., and will cut much faster than sand for certain classes of stone. In lieu of ordinary plain blades for cutting, some made with grooves or corrugations, with the object of conveying the cutting material more rapidly to the stone, have been introduced, but the author has so far had no opportunity of testing their merits.

FIG. 41.—CIRCULAR STONE SAWING MACHINE.   To face page 171.

CIRCULAR SAWS FOR CUTTING STONE.—A considerable number of these machines have lately been introduced, and for rapidly sawing, facing and edging building stone of a moderate degree of hardness, such as Portland, are found to be very expeditious.

We illustrate herewith a machine of recent construction from the designs of Mr. G. Anderson. The framing and gear is of massive construction to overcome the vibration of working. The table carrying the stone is traversed by a screw feed fitted with a quick return motion, its speed being varied to suit the nature of the stone. The gearing for driving the saws is of double purchase, and all the wheels and pinions are of strong pitch and heavily proportioned. The cutting teeth are adjustable, they are of the Hunter type, and can be readily made or renewed by an ordinary blacksmith. By mounting a disc on the spindle with the cutting tools projecting from its side, these machines can be adapted for rapidly facing as well as sawing the stone.

STONE-DRESSING AND PLANING MACHINES.—Some recent progress has been made with these machines, notably with those for dressing very hard stones, an illustration of which we give elsewhere. In machines for building stones based on the ordinary planing machine for iron, various modifications have been introduced, and most of them by automatically reversing the cutting tools are made to cut during both the backward and forward traverse of the table, and the cross-head carrying the tool box is raised and lowered by a belt. Supplementary canting tables are also fitted on the main table so that the stone can be dressed on three sides at any desired angle without re-setting. A rectilinear stone planing machine especially adapted for dressing flags, pavement coping, &c., has also been introduced;

this consists briefly of a heavy cast-iron slide, arranged to move on four turned rollers transversely across the face of the table carrying the stone. Four tools are employed, two roughing and two finishing, these are mounted in pairs in adjustable tool-holders, fixed on rocking pins at either side of the cast-iron slide or saddle. Both sets of tools are worked independently, and can be applied to different stones if desired. All the operations are automatic.

RUBBING BEDS.—In lieu of the ordinary type of machines, which require a pit and somewhat expensive foundations, the writer can recommend them being built self-contained, as they are much more readily fixed or removed, and the only foundation required being a bed of concrete. The frame of the machine should be massive, and the bed supported by friction rollers, as there is less trouble with the footstep bearing, at the same time the machine can be run at a higher rate of speed.

STONE OR MARBLE-TURNING LATHES.—In the most recent practice for turning granite, marble, &c., the lathe is made of very substantial construction, and by preference, fitted with double beds, and each bed fitted with a strong saddle and arranged with a compound slide rest, with a swivelling tool-holder for the circular revolving steel cutters. The slides should have hand adjustments and the saddle have a self-acting motion. The beds should be carried on sole plates, and so arranged that they can be set either parallel or to a taper of say 1in. in 4ft. The fast headstock should be fitted with a speed cone and double purchase spur gearing.

APPENDIX B.

TECHNICAL TERMS.

Architrave—is a term applied to the assemblage of members or mouldings which surround a door or window.

Archivolt—an ornamental moulding running round an arch.

Argillaceous shale or clay slate has the same constitution as Mica slate; but the particles are so fine as not to be distinguished.

Arris—the line of concourse, edge, or meeting of two surfaces.

Ashlar—when each stone is squared and dressed to given dimensions.
,, **Bastard**—is ashlar work, backed up with inferior work.
,, **Boasted**—same as chiselled.
,, **Chiselled**—a random tooled ashlar wrought with a narrow chisel.
,, **Herring-bone**—has a tooling of oblique flutes in ranks running in alternate directions.
,, **Nigged**—a building block dressed with a pointed hammer.
,, **Pick**, or hammer-dressed: it is known as common ashlar.
,, **Plane**—a block in which the marks of the tools are dressed out.
,, **Pointed**—the face marking done by a pointed tool or one very narrow.
,, **Prison**—the surface is wrought into holes.
,, **Random-tooled**—a block whose groovings are irregularly cut with a broad tool.
,, **Rough Ashlar**—a block of Freestone as brought from the quarry.
,, **Rusticated**—the face of the block projects from the joint, arrises being bevelled. It may be rough or smooth faced, or variously tooled.
,, **Smooth**—a block dressed ready for use.

Ashlar, Tooled—a block in which the surface has parallel vertical flutes.
> A smooth face around the joint is called a margin-draft. The face of an ashlar is the front exposed surface when built into the wall. Flanks, are the sides. Beds, are the upper and lower surfaces. Backs, are the rear surfaces.

Band—denotes any flat low member, or moulding that is broad and not very deep.

Bed-moulding—a collective term for all the moulding beneath the corona or principal projecting member of a cornice.

Bedding stone—a marble slab accurately level and on which pieces of stone or bricks are tried to prove the truth of their faces.

Beds of a stone—the upper and lower surfaces of a block.

Birdsbeak—a complete moulding used in the Greek Doric order.

Bituminous shale—is a slate clay impregnated with bitumen.

Blocking course—a course of masonry laid on the top of a cornice crowning a wall.

Boasted or chiselled—wrought by a narrow chisel.

Boaster—a stone-mason's chisel, having an edge two inches wide, used in dressing down the surface of stone. It is intermediate in width between the *inch tool* and the *broad tool*, which are respectively 1 inch and $3\frac{1}{2}$ inches wide.

Borer—a tool for boring rocks.

Boss—a sculptured keystone.

Boultine—a convex moulding whose periphery is a quarter of a circle next below the plinth in the Doric and Tuscan orders.

Boulder—made up of rounded unwrought stones.

Broached—hewed by mallet and chisel.

Cantilivers—are trusses when used under modillions in the frieze.

Chisel draught—a plane surface formed along the edge of a block of stone by the mason before commencing to dress the surface.

Chlorite slate—is of a dark green colour, similar to talcous slate.

Churn jumpers—so called from their method of working.

Claying bar—an implement for driving in dry clay into the hole made by the jumping tool, if found too damp for the introduction of the blasting material.

Cleaving grain—lines in the stone parallel to the planes of cleavage.

APPENDIX.

Console—is an ornament like a truss carried on a key-stone.

Compound mouldings—are the *cyma recta*, which has the hollow uppermost and projecting. The *cyma reversa*, which has a similar contour, but adapted for a base moulding to a wall or plinth.

Coursed work—in which the stones are squared more or less and set in courses.

Cramp or dowel—a piece of iron, copper, or wood usually of a dovetail form, used with lead for joining masonry. Dowels are sometimes made with hard stone, run with cement.

Cutter barrel, cutter head, cutter block, tool box. Apparatus for holding the working tools or cutters in stone-dressing machinery.

Dentils—are ornaments used in the bed moulds of cornices.

Dowel joggles—are usually hard pieces of stone cut and let into corresponding mortices in the two stones to be joined together.

Dressing—is the working on the faces and beds.

Dressings—the mouldings, and sculptured decorations of all kinds which are used on the walls and ceiling of a building for the purpose of ornament.

Droving work—is first droved and then broached.

Enriched mouldings—the above mouldings carved.

Feathers—are inverted wedges with circular backs.

Flutes—channels running down columns. These channels are sometimes partly filled by a lesser round moulding; this is called cabling the flutes.

Freestone—is applied indefinitely to that kind of stone which can readily be wrought. It includes the two great divisions of limestone and sandstone.

Grit—Coarse sand, rough hard particles of sandstone.

Ground table stones—the projecting course of stones in a wall above the plinth.

Hammer dressed—squared and picked by the hammer.

Herring-bone—when the surfaces of the stone are formed with zig-zag lines.

Hornblende slate—resembles mica-slate, but is less glistening and does not break into such thin sheets.

Impost—a horizontal moulding for an arch to spring from.

Joggled stones—are jointed together when a projection is worked out on one stone to fit into a corresponding hole or groove in the other.

Jumping tool—implement for boring.

Key-stone—the stone in an arch which is equally distant from its springing extremities.

Lintel or traverse—the cross top.

Micaceous—stones laminated with mica.

Mica slate—a rock composed of quartz and mica.

Modillions—small brackets under the corona in the cornice.

Mouldings—may be divided into two classes, the simple and the compound ; the former are :—
1st—The Roman *ovolo*, or quarter round ; or the Greek *ovolo*, with an elliptic or conic section.
2nd—The *cavetto* or hollow.
3rd—The *torus* or round.

Nigged work—is that in which the work is picked with a scabbling hammer until the surface is nearly of the intended form.

Ogee—a moulding which has the round uppermost and overhanging.

Oolites or roestones—are composed of oviform bodies cemented by calcareous matter of a varied character.

Ovolo—a convex moulding mostly used in classical architecture.

Parpoint—squared stones laid in courses, with courses of *headers* at intervals of, say, five feet.

Plane ashlar—rubbed smooth.

Planes of clearage—stone which occurs in contiguous strata presents a number of contact surfaces called planes of clearage.

Plucked—a dressed stone with an uneven and holey surface is plucked.

Point, inch tool, boaster, broad tool—are mason's tools usually employed for dressing the face of stones.

Reedings—are several beads placed together.

Rough—axed on the face.

Rubbed work—when the surface of the stone is smoothed by means of sand or grit stone.

Rubble—a filling in with irregular stone.

Rubble work—in which the stones are not squared.

Rustic masonry—if the joints of the masonry are sunk in channels the work is called rustic.

Scabbling—dressing the surface of the stone in the quarry with pointed picks.

Scotia—is formed of two hollows, one over the other, and of different centres.

Scotia—the hollow moulding in the base of an Ionic column.

Scraper—a tool for clearing the hole of stone chips and dust made by the boring tool.

Slate adhesive—of a light greenish grey colour, is readily broken, and adheres to the tongue.

Soffit—is the under part of a lintel or cornice.

Spaulled—the action of splitting or scaling off small flakes of stone.

Stone axe, Jedding axe, Scabbling hammer or cavil—is used previous to the operation of hewing, in order to bring the stone nearly to shape; one end of the jedding-axe is flat, and is used for knocking off projecting angular points, and the other end is pointed for reducing the different surfaces nearly to the intended form.

Stroking—dressing stone by successive narrow flutes or grooves.

Talcose slate—contains talc instead of mica, has a more greasy feel than mica-slate.

Tamping-bar—a tool for filling up a bore-hole made for blasting.

Through-stones—or bond stones, are those placed with their greatest length going through the thickness of the wall at a right angle to its face.

Truss—is a modillion enlarged and placed flat against a wall.

Unstunned—even on the face, not "plucked."

Weather moulding—a dripstone over a door or window to keep off water from the parts beneath.

INDEX.

ARR

ARRANGEMENT of shafting, 152
— — stone-works, 16
Art of stone-working, origin of, 1

BAND-saws with diamond points, 36
Bearings, lubricant for, 19
Bed, old moulding, 77
— — ripping, 41
— polishing, 109
— rubbing or surfacing, 89
Bedding, stone, 14

CARVING machinery, 103
Chairing and boring machine for railway blocks, 135
Checking deep mouldings, 85
Circular, revolving, or rolling cutters for dressing stone, 63
Circular saws, construction of, 50
— — for cutting stone, 40, 171
— — for roughing out mouldings, 100
— — hydraulic feed for, 127
— — speed of, 49, 161
— — teeth of, 42
Construction of stone-sawing frames, 31
Cost of stone-working by machinery, 155
Cutter blocks, moulding, composition of, 99
Cutters, complex forms to be avoided, 85
— rolling for turning, 117
— tempering, 85, 139
Cutting difficult stone, 49
— marble, 34
— speed of circular-saws, 161
— tools, 138

GUI

Cutting window sills, coping-stones, &c., 49
Cylindrical pillars, method of cutting, 26

DECOMPOSITION of stone, 12
Diamond drill, 38
Diamond points for sawing, 36
Disc surfacing machines, 75
Disintegration of stone, 13
Dressing and planing machines for stone, 52
— granite, 75
— millstones, 136
Driving belts, 152
Duplex stone-moulding machine, 81

EDGE moulding, 87
Emery wheels, speed of, 145

FACING stone of awkward section, 94
First machine for dressing stone, 54
Foundation of rubbing bed, 93
Frame, saw, for marble, 33
Frames, stone-sawing, 22
— — — construction of, 31
— — — guides for, 32
— — — skeleton, 30
— wooden, for holding stone, 84

GRANITE dressing, 75
— its composition, 8
— turning, 116
Grinding and polishing materials, 110
— moulding irons, 144
Grindstone dressing machine, 137
Guides for saw frames, 32

HAND rubbers for polishing, 113
— Hard stone dressing with diamond points, 37
Head stock for holding pillars, 84
Holing slates, 131
Horizontal saw frame, 170
Hydraulic feed circular saw, 127

IRREGULAR moulding machine, 99

LABOUR, hand, *versus* machinery, 155
Lathe for stone-turning, 115
Limestone, its composition, 8
Lubricants for bearings, 19

MACHINE conversion, stone suitable for, 8
Machine, first, for dressing stone, 54
Machinery for breaking stone, 118
— — dressing slate, 126
— — — millstones, 136
— — planing slate, 131
— — working slate, 124
— hand labour *versus*, 155
— miscellaneous, for working stone, 135
Management of stone-works, 149
Marble cutting, 34
— quarrying, 135
— saw frame for, 33
— white, 164
Masonry, Notes on, 162
Materials for grinding and polishing, 10
Motive power for stone-works, 19
Moulding irons, grinding, 144
Moulding machines, classes of, 79
— — edge, 87
— — for stone, 77
— — horizontal barrel, 86
— — irregular, 99
Mouldings, "checking" deep, 85

PILLARS, cylindrical, method of cutting, 26
— headstock for holding, 84
Plan of stone-works for general purposes, 17
Planing machines for stone, 52
Polishing beds, 109
— hand-rubbers for, 113
— marble, 33, 112

Polishing materials, 110
— stone of awkward section, 94

QUARRYING machinery, 100, 131, 135

RECESSING machinery, 98
— Recipes, 162
Rolling cutters for turning, 117
Rubbing or surfacing beds, 89, 172
— — — construction of, 93
— — — cost of, 92
— — — foundations of, 93

SAND and water, supplying, 32
— for sawing, 34
Sandstones, their composition, 8
Saws, circular, construction of, 50
— — for cutting stone, 40
— — teeth of, 42
Scraping tools for mouldings, 82, 83
Sculpturing machinery, 103
Shafting, arrangement of, 152
Slate dressing machine, 126
— holing machine, 131
— pencils, manufacture of, 129
— planing machine, 131
— Welsh, 165
— working machinery, 124
Speed of circular saw, 161
— — emery wheels, 145
— — table for moulding, 87
Stone, Bath, 165
— breaking machinery, 118
— the bedding of, 14
— conversion of, to best advantage, 24
— Cragleith, 165
— decomposition of, 12
— difficult cutting, 49
— disintegration of, 13
— dressing and planing machines, 52, 171
— Dundee, 165
— found in Great Britain, 14
— grooving machine, 100
— moulding machines, 77
— polishing of awkward section, 94
— Portland, 165
— preservation of, 168
— recessing, 98
— sawing frames, 22, 30, 33
— — — construction of, 31
— — — guide for, 32
— — — machinery, introduction of, 25

STO

Stone Sawing, the term a misnomer, 22
— — with diamond points, 36
— saws, circular, 40
— seasoning, 11
— suitable for machine conversion, 8
— surfacing or rubbing beds, 89
— testing, 11
— turning lathe, 115, 172
— the hardest not always worst to work, 9
Stone-working, hand labour *versus* machinery, 155
— art of, supposed origin of, 1
— machinery, miscellaneous, 135
Stoneworks, arrangement of, 16
— management of, 149
— motive power for, 19

WIN

TECHNICAL terms, 173
   Tempering cutters, 85, 139
Tools, Cook & Hunter's Patent, 43
— cutting, 138
— for moulding machine, 80, 81
— reversible, 73
— revolving, 53, 63
— with diamond points, 37
Turning lathes, 115
— granite, 116

VERTICAL moulding machines, 79, 87

WINDOW sills, coping stones, &c. cutting, 49

M. POWIS BALE & Co.

(Formerly with the late Firm, POWIS, JAMES, & CO.*),*

ENGINEERS,

OFFICES AND WORKSHOPS: APPOLD STREET, LONDON, E.C.

SPECIALITIES:

WOODWORKING MACHINERY,
STONEWORKING MACHINERY,
STEAM ENGINES, &c.,

OF THE MOST ADVANCED CONSTRUCTION.

ALSO

PUMPING AND COLONIAL MACHINERY GENERALLY.

MILLS PLANNED AND ERECTED COMPLETE.

ENGINEERING VALUATIONS, ARBITRATIONS, EXPERT EVIDENCE, &c.

BOOKS BY M. POWIS BALE, M.Inst.M.E., A.M.Inst.C.E.

Now Published:

"WOODWORKING MACHINERY, ITS RISE, PROGRESS, AND CONSTRUCTION." 9s.

"SAW-MILLS, THEIR ARRANGEMENT AND MANAGEMENT." 10s. 6d.

"STONEWORKING MACHINERY, AND THE RAPID AND ECONOMICAL CONVERSION OF STONE." 9s.

"STEAM AND MACHINERY MANAGEMENT." 2s. 6d.

"PUMPS AND PUMPING." 2s. 6d.

"A HANDBOOK FOR STEAM USERS." 2s. 6d.

"MODERN SHAFTING AND GEARING." 2s. 6d.

ROSBY LOCKWOOD & SON'S

LIST OF WORKS

ON

CIVIL, MECHANICAL, ARINE AND ELECTRICAL ENGINEERING.

> A Complete Catalogue of NEW and STANDARD WORKS on MINING and COLLIERY WORKING; ARCHITECTURE and BUILDING; The INDUSTRIAL ARTS, TRADES and MANUFACTURES; CHEMISTRY and CHEMICAL MANUFACTURES; AGRICULTURE, FARMING, GARDENING, AUCTIONEERING, LAND AGENCY, &c. Post Free on Application.

7, STATIONERS' HALL COURT, LONDON, E.C.
1907.

LIST OF WORKS

ON

CIVIL, MECHANICAL, ELECTRICAL

AND MARINE ENGINEERING.

AËRIAL NAVIGATION. A Practical Handbook on the Construction of Dirigible Balloons, Aërostats, Aëroplanes, and Aëromotors. By FREDERICK WALKER, C.E., Associate Member of the Aëronautic Institute. With 104 Illustrations. Large Crown 8vo, cloth . . . *Net* **7/6**

AËRIAL OR WIRE-ROPE TRAMWAYS. Their Construction and Management. By A. J. WALLIS-TAYLER, A.M.Inst.C.E. With 81 Illustrations. Crown 8vo, cloth **7/6**

ARMATURE WINDINGS OF DIRECT CURRENT DYNAMOS. Extension and Application of a General Winding Rule. By E. ARNOLD, Engineer, Assistant Professor in Electro-Technics and Machine Design at the Riga Polytechnic School. Translated from the original German by FRANCIS B. DE GRESS, M.E., Chief of Testing Department, Crocker-Wheeler Company. Medium 8vo, 120 pp., with over 140 Illustrations *Net* **12/0**

BEAMS. EXPERIMENTS ON THEIR FLEXURE. Resulting in the Discovery of New Laws of Failure by Buckling. By ALBERT E. GUY. Medium 8vo, cloth . . . . *Net* **9/0**

BLAST FURNACE CALCULATIONS AND TABLES FOR FURNACE MANAGERS AND ENGINEERS. Containing Rules and Formulæ for Finding the Dimensions and Output Capacity of any Furnace, as well as the regular Outfit of Stoves, Heating Surface, Volume of Air, Tuyere Area, &c., per ton of Iron per day of 24 hours. By JOHN L. STEVENSON. F'cap. 8vo, leather. . . . . *Net* **5/0**

BOILER AND FACTORY CHIMNEYS. Their Draught-Power and Stability. With a chapter on "Lightning Conductors." By ROBERT WILSON, A.I.C.E., Author of "A Treatise on Steam Boilers," &c. Crown 8vo, cloth **3/6**

BOILER CONSTRUCTION. A Practical Handbook for Engineers, Boiler-Makers, and Steam Users. Containing a large Collection of Rules and Data relating to Recent Practice in the Design, Construction, and Working of all Kinds of Stationary, Locomotive, and Marine Steam-Boilers. By WALTER S. HUTTON, Civil and Mechanical Engineer. With upwards of 500 Illustrations. Fourth Edition, carefully Revised, and Enlarged. Medium 8vo, over 680 pages, cloth, strongly bound **18/0**

HEAT, RADIATION, AND CONDUCTION—NON-CONDUCTING MATERIALS AND COVERINGS FOR STEAM-BOILERS—COMPOSITION, CALORIFIC-POWER AND EVAPORATIVE-POWER OF FUELS—COMBUSTION, FIRING STEAM-BOILERS, PRODUCTS OF COMBUSTION, ETC., CHIMNEYS FOR STEAM-BOILERS—STEAM BLAST—FORCED DRAUGHT—FEED-WATER—EFFECT OF HEAT ON WATER—EXPANSION OF WATER BY HEAT—WEIGHT OF WATER AT DIFFERENT TEMPERATURES—CONVECTION—CIRCULATION—EVAPORATION—PROPERTIES OF SATURATED STEAM—EVAPORATIVE POWER OF BOILERS—PRIMING, ETC.—WATER-HEATING SURFACES OF STEAM-BOILERS—TRANSMISSION OF HEAT—SMOKE-TUBES—EVAPORATIVE POWER AND EFFICIENCY OF BOILERS—WATER-CAPACITY AND STEAM-CAPACITY OF BOILERS—FIRE-GRATES, FIRE BRIDGES, AND FIRE-BARS—POWER OF BOILERS—CYLINDRICAL SHELLS AND FURNACE-TUBES OF BOILERS, ETC.—TESTS OF MATERIALS—STRENGTH AND WEIGHT OF BOILER-PLATES—EFFECT OF TEMPERATURE ON METALS—RIVET HOLES—RIVETS—RIVETED JOINTS OF STEAM-BOILERS—CAULKING ENDS OF CYLINDRICAL-SHELLS—STAYS FOR BOILERS, ETC.—STEAM GENERATORS—DESCRIPTION AND PROPORTIONS OF CORNISH, LANCASHIRE, AND OTHER TYPES OF STATIONARY BOILERS—BOILER-SETTING—MULTI-TUBULAR, LOCOMOTIVE, PORTABLE, MARINE, VERTICAL, AND WATER-TUBE BOILERS—SUPER-HEATERS—COST OF STEAM PRODUCTION—FURNACES FOR REFUSE-FUELS—DESTRUCTORS, ETC.—SAFETY-VALVES—STEAM PIPES—STOP-VALVES AND OTHER MOUNTINGS FOR BOILERS—FEED-PUMPS—STEAM PUMPS—FEED-WATER CONSUMPTION—INJECTORS—INCRUSTATION AND CORROSION—FEED-WATER HEATERS—EVAPORATORS—TESTING BOILERS—EVAPORATIVE PERFORMANCES OF STEAM BOILERS; STEAM-BOILER EXPLOSIONS, ETC.

"There has long been room for a modern handbook on steam boilers; there is not that room now, because Mr. Hutton has filled it. It is a thoroughly practical book for those who are occupied in the construction, design, selection, or use of boilers."—*Engineer.*

BOILERMAKER'S ASSISTANT. In Drawing, Templating, and Calculating Boiler Work, &c. By J. COURTNEY, Practical Boilermaker. Edited by D. K. CLARK, C.E. Seventh Edition. Crown 8vo, cloth . **2/0**

BOILERMAKER'S READY RECKONER. With Examples of Practical Geometry and Templating for the Use of Platers, Smiths, and Riveters. By John COURTNEY. Edited by D. K. CLARK, M.Inst.C.E. Crown 8vo, cloth **4/0**

BOILERMAKER'S READY RECKONER & ASSISTANT, being the two previous mentioned volumes bound together in one volume. With Examples of Practical Geometry and Templating, for the Use of Platers, Smiths, and Riveters. By JOHN COURTNEY. Edited by D. K. CLARK, M.Inst.C.E. Fifth Edition, 480 pp., with 140 Illustrations. Crown 8vo, half bound **7/0**

"No workman or apprentice should be without this book."—*Iron Trade Circular.*

BOILER MAKING AND PLATING. A Practical Handbook for Workshop Operations. By JOSEPH G. HORNER, A.M.I.M.E. 380 pp. with 338 Illustrations. Crown 8vo, cloth **7/6**

BOILERS (STEAM). Their Construction and Management. By R. ARMSTRONG, C.E. Illustrated. Crown 8vo, cloth . . . **1/6**

BOILERS. Their Strength, Construction, and Economical Working. By R. WILSON, C.E. Fifth Edition. 12mo, cloth . **6/0**

CIVIL, MECHANICAL, &c., ENGINEERING. 5

BRIDGE CONSTRUCTION IN CAST AND WROUGHT IRON. A Complete and Practical Treatise on, including Iron Foundations. In Three Parts.—Theoretical, Practical, and Descriptive. By WILLIAM HUMBER A.M.Inst.C.E., and M.Inst.M.E. Third Edition, revised and much improved, with 115 Double Plates (20 of which now first appear in this edition), and numerous Additions to the Text. In 2 vols., imp. 4to, half-bound in morocco . . . . **£6 16s. 6d.**

"In addition to elevations, plans, and sections, large scale details are given, which very much enhance the instructive work of those illustrations."—*Civil Engineer and Architect's Journal.*

BRIDGES AND VIADUCTS, IRON AND STEEL. A Practical Treatise upon their Construction. For the use of Engineers, Draughtsmen, and Students. By FRANCIS CAMPIN, C.E. Crown 8vo, cloth **3/6**

BRIDGES (IRON) OF MODERATE SPAN: Their Construction and Erection. By H. W. PENDRED. With 40 illustrations. Crown 8vo, cloth **2/0**

BRIDGES, OBLIQUE. A Practical and Theoretical Essay. With 13 large Plates. By the late GEORGE WATSON BUCK, M.Inst.C.E. Fourth Edition, revised by his Son, J. H. WATSON BUCK, M.Inst.C.E.; and with the addition of Description to Diagrams for Facilitating the Construction of Oblique Bridges, by W. H. BARLOW, M.Inst.C.E. Royal 8vo, cloth **12/0**

"As a guide to the engineer and architect, on a confessedly difficult subject, Mr. Buck's work is unsurpassed."—*Building News.*

BRIDGES, TUBULAR AND OTHER IRON GIRDER. Describing the Britannia and Conway Tubular Bridges. With a Sketch of Iron Bridges, &c. By G. D. DEMPSEY, C.E. Crown 8vo, cloth . **2/0**

CALCULATOR (NUMBER, WEIGHT, AND FRACTIONAL). Containing upwards of 250,000 Separate Calculations, showing at a Glance the Value at 422 Different Rates, ranging from $\frac{1}{16}$th of a Penny to 20s. each, or per cwt., and £20 per ton, of any number of articles consecutively, from 1 to 470. Any number of cwts., qrs., and lbs., from 1 cwt. to 470 cwts. Any number of tons, cwts., qrs., and lbs., from 1 to 1,000 tons. By WILLIAM CHADWICK, Public Accountant. Fourth Edition, Revised and Improved. 8vo, strongly bound **18/0**

CALCULATOR (WEIGHT). Being a Series of Tables upon a New and Comprehensive Plan, exhibiting at one Reference the exact Value of any Weight from 1 lb. to 15 tons, at 300 Progressive Rates, from 1d. to 168s. per cwt., and containing 186,000 Direct Answers, which, with their Combinations, consisting of a single addition (mostly to be performed at sight), will afford an aggregate of 10,266,000 Answers; the whole being calculated and designed to ensure correctness and promote despatch. By HENRY HARBEN, Accountant. Sixth Edition, carefully Corrected. Royal 8vo, strongly half-bound . . . . **£1 5s.**

CHAIN CABLES AND CHAINS. Comprising Sizes and Curves of Links, Studs, etc., Iron for Cables and Chains Chain Cable and Chain Making, Forming and Welding Links, Strength of Cables and Chains, Certificates for Cables, Marking Cables, Prices of Chain Cables and Chains, Historical Notes, Acts of Parliament, Statutory Tests. Charges for Testing, List of Manufacturers of Cables, etc., etc. By THOMAS W. TRAILL, F.E.R.N., M.Inst.C.E., Engineer-Surveyor-in-Chief, Board of Trade, Inspector of Chain Cable and Anchor Proving Establishments, and General Superintendent, Lloyd's Committee on Proving Establishments. With numerous Tables, Illustrations, and Lithographic Drawings. Folio, cloth **£2 2s.**

CIVIL ENGINEERING. By HENRY LAW, M.Inst.C.E. Including a Treatise on Hydraulic Engineering by G. R. BURNELL, M.Inst.C.E. Seventh Edition, revised, with Large Additions on Recent Practice by D. KINNEAR CLARK, M.Inst.C.E. Crown 8vo, cloth **6/6**

CONDUCTORS FOR ELECTRICAL DISTRIBUTION. Their Materials and Manufacture, The Calculation of Circuits, Pole-Line Construction, Underground Working, and other Uses. By F. A. C. PERRINE, A.M., D.Sc.; formerly Professor of Electrical Engineering, Leland Stanford, Jr., University; M.Amer.I.E.E. Medium 8vo, 300 pp., fully illustrated, including Folding Plates and Diagrams . . . . . *Net* **20/0**

CONTINUOUS RAILWAY BRAKES. A Practical Treatise on the several Systems in Use in the United Kingdom, their Construction and Performance. By M. REYNOLDS. 8vo, cloth **9/0**

CRANES, the Construction of, and other Machinery for Raising Heavy Bodies for the Erection of Buildings, &c. By J. GLYNN, F.R.S. Crown 8vo, cloth **1/6**

CURVES, TABLES OF TANGENTIAL ANGLES AND MULTIPLES. For Setting-out Curves from 5 to 200 Radius. By A. BEAZELEY, M.Inst.C.E. 7th Edition, Revised. With an Appendix on the use of the Tables for Measuring up Curves: Printed on 50 Cards, and sold in a cloth box, waistcoat-pocket size **3/6**

"Each table is printed on a small card, which, placed on the theodolite, leaves the hands free to manipulate the instrument—no small advantage as regards the rapidity of work."—*Engineer.*

DRAINAGE OF LANDS, TOWNS AND BUILDINGS. By G. D. DEMPSEY, C.E. Revised, with large Additions on Recent Practice in Drainage Engineering by D. KINNEAR CLARK, M.Inst.C.E. Fourth Edition. Crown 8vo, cloth **4/6**

DYNAMIC ELECTRICITY AND MAGNETISM, ELEMENTS OF. A Handbook for Students and Electrical Engineers. By PHILIP ATKINSON, A.M., Ph.D. Crown 8vo, cloth, 417 pp., with 120 Illustrations **10/6**

DYNAMO BUILDING. HOW TO MAKE A DYNAMO. A Practical Treatise for Amateurs. By ALFRED CROFTS. Crown 8vo, cloth **2/0**

DYNAMO CONSTRUCTION. A Practical Handbook for the Use of Engineer-Constructors and Electricians-in-Charge. By J. W. URQUHART. Crown 8vo, cloth **7/6**

DYNAMO ELECTRIC MACHINERY. By SAMUEL SHELDON, A.M., Ph.D., Professor of Physics and Electrical Engineering at the Polytechnic Institute of Brooklyn, etc., assisted by HOBART MASON, B.S., E.E.

In two volumes (sold separately).

Vol. I.—DIRECT CURRENT MACHINES. Sixth Edition, Revised. 202 Illustrations . . . . *Net* **12/0**

Vol. II.—ALTERNATING CURRENT MACHINES. Fifth Edition. With 184 Illustrations . . . . *Net* **12/0**

DYNAMO MANAGEMENT. A Handybook of Theory and Practice for the Use of Mechanics, Engineers, Students, and others in Charge of Dynamos. By G. W. LUMMIS-PATERSON, Electrical Engineer. Third Edition, Revised. Crown 8vo, 260 pp., with 100 illustrations, cloth . **4/6**

ELECTRICAL UNITS—MAGNETIC PRINCIPLES—THEORY OF THE DYNAMO—ARMATURES—ARMATURES IN PRACTICE—FIELD MAGNETS—FIELD MAGNETS IN PRACTICE—REGULATION DYNAMOS — COUPLING DYNAMOS — RUNNING DYNAMOS — FAULTS IN DYNAMOS—FAULTS IN ARMATURES—MOTORS.

"The book may be confidently recommended."—*Engineer.*

DYNAMO, MOTOR AND SWITCHBOARD CIRCUITS FOR ELECTRICAL ENGINEERS. A Practical Book dealing with the subject of Direct, Alternating and Polyphase Currents. By WILLIAM R. BOWKER, C.E., M.E., E.E. Medium 8vo, cloth. 109 Diagrams *Net* **6/0**

DYNAMO AND MOTOR CIRCUITS—STARTING AND STOPPING OF SAME—METHODS OF CHANGING DIRECTION OF ROTATION—SYNCHRONISM—PARALLELING OF ALTERNATORS, ETC.—POLYPHASE CIRCUITS—POLYPHASE TRANSMISSION OF POWER—DIPHASE AND TRIPHASE CIRCUITS, ETC. — BOOSTERS — EQUALISERS — REVERSIBLE BOOSTERS—STORAGE BATTERIES—END-CELL SWITCHES, ETC.—ELECTRIC TRACTION MOTORS — SERIES — PARALLEL CONTROLLERS — CAR WIRING DIAGRAMS — MOTOR VEHICLE CIRCUITS — CANAL HAULAGE ROTARY CONVERTERS — SWITCHBOARD CIRCUITS, ETC.

DYNAMOS (ALTERNATING AND DIRECT CURRENT). A Text-book on their Construction for Students, Engineer-Constructors and Electricians-in-Charge. By TYSON SEWELL, A.M.I.E.E., Lecturer and Demonstrator in Electrical Engineering at the Polytechnic, Regent Street, London, author of "The Elements of Electrical Engineering." 328 pp., with over 230 Illustrations. Large crown 8vo, cloth. [*Just published.* *Net* **7/6**

FUNDAMENTAL PRINCIPLES OF DIRECT CURRENTS—THE MAGNETIC FIELD—THE PRODUCTION OF AN ELECTRO-MOTIVE FORCE—FUNDAMENTAL PRINCIPLES OF ALTERNATING CURRENTS—THE ALTERNATING MAGNETIC FIELD—THE CAPACITY OF THE CIRCUIT—BIPOLAR DYNAMO CONSTRUCTION—THEORY OF BIPOLAR MACHINES—BIPOLAR DYNAMO DESIGN — MULTIPOLAR DYNAMO CONSTRUCTION — MULTIPOLAR DYNAMO DESIGN—SINGLE PHASE ALTERNATORS—CONSTRUCTION OF ALTERNATORS—POLYPHASE ALTERNATORS — EXCITING, COMPOUNDING AND SYNCHRONISING OF ALTERNATORS.

EARTHWORK MANUAL. By ALEX. J. GRAHAM, C.E. With numerous Diagrams. Second Edition. 18mo, cloth **2/6**

EARTHWORK TABLES. Showing the Contents in Cubic Yards of Embankments, Cuttings, &c., of Heights or Depths up to an average of 80 feet By JOSEPH BROADBENT, C.E., and FRANCIS CAMPIN, C.E. Crown 8vo, cloth **5/0**

EARTHWORK TABLES, HANDY GENERAL. Giving the Contents in Cubic Yards of Centre and Slopes of Cuttings and Embankments from 3 inches to 80 feet in Depth or Height, for use with either 66 feet Chain or 100 feet Chain. By J. H. WATSON BUCK, M.Inst.C.E. On a Sheet mounted in cloth case **3/6**

ELECTRIC LIGHT. Its Production and Use. By J. W. URQUHART. Crown 8vo, cloth **7/6**

ELECTRIC LIGHT FITTING. A Handbook for Working Electrical Engineers. By J. W. URQUHART. Crown 8vo, cloth . . **5/0**

ELECTRIC LIGHT FOR COUNTRY HOUSES. A Practical Handbook, including Particulars of the Cost of Plant, and Working. By J. H. KNIGHT. Crown 8vo, wrapper **1/0**

ELECTRIC LIGHTING. By ALAN A. CAMPBELL SWINTON, M.Inst.C.E., M.I.E.E. Crown 8vo, cloth **1/6**

ELECTRIC LIGHTING AND HEATING POCKET BOOK. Comprising useful Formulæ, Tables, Data, and Particulars of Apparatus and Appliances for the use of Central Stations Engineers, Contractors, and Engineers-in-Charge. By SYDNEY F. WALKER, R.N., M.I.E.E., M I.M.E., A.M.Inst.C.E., Etc. F'cap 8vo, 466 pp., 270 Diagrams, and 240 Tables. *[Just published.* Net **7/6**

DEFINITIONS—DIFFERENT UNITS EMPLOYED—LAWS OF ELECTRIC CIRCUITS—DIFFERENCES BETWEEN WORKING OF CONTINUOUS AND ALTERNATING CURRENTS—LAWS OF ELECTRO MAGNETIC AND ELECTRO - STATIC INDUCTION — ELECTRICITY GENERATORS—ACCUMULATORS—SWITCHBOARDS—SWITCHES, CIRCUIT-BREAKERS, ETC. —CABLES—METHODS OF INSULATION—SIZES AND INSULATION OF CABLES MADE BY LEADING MAKERS—CONDUITS—LEADING WIRES AND OTHER ACCESSORIES—MEASURING INSTRUMENTS OF ALL KINDS AND APPARATUS FOR TESTING LAMPS AND ACCESSORIES—APPARATUS FOR HEATING BY ELECTRICITY.

ELECTRIC SHIP-LIGHTING. A Handbook on the Practical Fitting and Running of Ships' Electrical Plant. By J. W. URQUHART. Crown 8vo, cloth **7/6**

ELECTRIC-WIRING, DIAGRAMS & SWITCH-BOARDS. By NEWTON HARRISON, E.E., Instructor of Electrical Engineering in the Newark Technical School. Crown 8vo, cloth. . . . *Net* **5/0**

THE BEGINNING OF WIRING—CALCULATING THE SIZE OF WIRE—A SIMPLE ELECTRIC LIGHT CIRCUIT CALCULATED—ESTIMATING THE MAINS, FEEDERS, AND BRANCHES—USING THE BRIDGE FOR TESTING—THE INSULATION RESISTANCE—WIRING FOR MOTORS—WIRING WITH CLEATS, MOULDING AND CONDUIT—LAYING-OUT A CONDUIT SYSTEM—POWER REQUIRED FOR LAMPS—LIGHTING OF A ROOM—SWITCHBOARDS AND THEIR PURPOSE—SWITCHBOARDS DESIGNED FOR SHUNT AND COMPOUND-WOUND DYNAMOS—PANEL SWITCHBOARDS, STREET RAILWAY SWITCHBOARDS, LIGHTNING ARRESTERS—THE GROUND DETECTOR—LOCATING GROUNDS—ALTERNATING CURRENT CIRCUITS—THE POWER FACTOR IN CIRCUITS—CALCULATION OF SIZES OF WIRE FOR SINGLE, TWO AND THREE-PHASE CIRCUITS.

ELECTRICAL AND MAGNETIC CALCULATIONS. For the Use of Electrical Engineers and Artisans, Teachers, Students, and all others interested in the Theory and Application of Electricity and Magnetism. By A. A. ATKINSON, M.S., Professor of Physics and Electricity in Ohio University, Athens, Ohio. Crown 8vo, cloth . . . . *Net* **9/0**

ELECTRICAL DICTIONARY. A Popular Encyclopædia of Words and Terms Used in the Practice of Electrical Engineering. By T. O'CONOR SLOANE, A.M., E.M., Ph.D., Author of "Arithmetic of Electricity," "Electricity Simplified," "Electric Toy Making," etc. Third Edition. with Appendix. 690 pages and nearly 400 Illustrations. Large Crown 8vo, cloth
Net **7/6**

"The work has many attractive features in it, and is, beyond doubt, a well put together and useful publication. The amount of ground covered may be gathered from the fact that in the index about 5,000 references will be found."—*Electrical Review.*

ELECTRICAL ENGINEERING. A First-Year's Course for Students. By TYSON SEWELL, A.I.E.E., Assistant Lecturer and Demonstrator in Electrical Engineering at the Polytechnic, Regent Street, London. Third Edition, Revised. with additional Chapters on Alternate Current Working and an Appendix of Questions and Answers. Large Crown 8vo, cloth. 442 pp., with 274 Illustrations . . . . . *Net* **7/6**

OHM'S LAW—UNITS EMPLOYED IN ELECTRICAL ENGINEERING—SERIES AND PARALLEL CIRCUITS; CURRENT DENSITY AND POTENTIAL DROP IN THE CIRCUIT—THE HEATING EFFECT OF THE ELECTRIC CURRENT—THE MAGNETIC EFFECT OF AN ELECTRIC CURRENT.—THE MAGNETISATION OF IRON.—ELECTRO-CHEMISTRY; PRIMARY BATTERIES.—ACCUMULATORS.—INDICATING INSTRUMENTS.—AMMETERS, VOLTMETERS, OHMMETERS—ELECTRICITY SUPPLY METERS—MEASURING INSTRUMENTS, AND THE MEASUREMENT OF ELECTRICAL RESISTANCE — MEASUREMENT OF POTENTIAL DIFFERENCE, CAPACITY, CURRENT STRENGTH, AND PERMEABILITY—ARC LAMPS—INCANDESCENT LAMPS, MANUFACTURE AND INSTALLATION; PHOTOMETRY—THE CONTINUOUS CURRENT DYNAMO—DIRECT CURRENT MOTORS—ALTERNATING CURRENTS—TRANSFORMERS, ALTERNATORS, SYNCHRONOUS MOTORS—POLYPHASE WORKING—APPENDIX OF QUESTIONS AND ANSWERS.

"Distinctly one of the best books for those commencing the study of electrical engineering. Everything is explained in simple language which even a beginner cannot fail to understand."—*The Engineer.*

ELECTRICAL ENGINEERING (ELEMENTARY). In Theory and Practice. A Class Book for Junior and Senior Students and Working Electricians. By J. H. ALEXANDER. With nearly 200 Illustrations. Crown 8vo, cloth . . . . *Net* **3/6**

FUNDAMANTAL PRINCIPLES—ELECTRICAL CURRENTS—SOLENOID COILS—GALVONOMETERS—VOLT-METERS—MEASURING INSTRUMENTS—ALTERNATING CURRENTS—DYNAMO ELECTRIC MACHINES—CONTINUOUS CURRENT DYNAMOS—INDUCTION, STATIC TRANSFORMERS, CONVERTERS—MOTORS—PRIMARY AND STORAGE CELLS—ARC LAMPS—INCANDESCENT LAMPS—SWITCHES, FUSES, ETC.—CONDUCTORS AND CABLES—ELECTRICAL ENERGY METERS—SPECIFICATIONS—GENERATION AND TRANSMISSION OF ELECTRICAL ENERGY—GENERATING STATIONS.

ELECTRICAL TRANSMISSION OF ENERGY. A Manual for the Design of Electrical Circuits. By ARTHUR VAUGHAN ABBOTT, C.E., Member American Institute of Electrical Engineers, Member American Institute of Mining Engineers, Member American Society of Civil Engineers, Member American Society of Mechanical Engineers &c. Fourth Edition, entirely Re-Written and Enlarged, with numerous Tables, 16 Plates, and nearly 400 other Illustrations. Royal 8vo, 700 pages. Strongly bound in cloth . . . . *Net* **30/0**

INTRODUCTION—THE PROPERTIES OF WIRE—THE CONSTRUCTION OF AERIAL CIRCUITS—GENERAL LINE WORK—ELECTRIC RAILWAY CIRCUITS—PROTECTION—THE CONSTRUCTION OF UNDERGROUND CIRCUITS—CONDUITS—CABLES AND CONDUIT CONDUCTORS—SPECIAL RAILWAY CIRCUIT—THE INTERURBAN TRANSMISSION LINE—THE THIRD RAIL—THE URBAN CONDUIT—ELECTRICAL INSTRUMENTS—METHODS OF ELECTRICAL MEASUREMENT — CONTINUOUS-CURRENT CONDUCTORS — THE HEATING OF CONDUCTORS—CONDUCTORS FOR ALTERNATING CURRENTS—SERIES DISTRIBUTION—PARALLEL DISTRIBUTION—MISCELLANEOUS METHODS—POLYPHASE TRANSMISSION—THE COST OF PRODUCTION AND DISTRIBUTION.

NOTE.—This Volume forms an indispensable Work for Electrical Engineers, Railway and Tramway Managers and Dirrectors, and all interested in Electric Traction.

ELECTRICITY AS APPLIED TO MINING. By Arnold Lupton, M.Inst.C.E., M.I.Mech.E., M.I.E.E., late Professor of Coal Mining at the Yorkshire College, Victoria University, Mining Engineer and Colliery Manager; G. D. Aspinall Parr, M.I.E.E., A.M.I.Mech.E., Associate of the Central Technical College, City and Guilds of London, Head of the Electrical Engineering Department, Yorkshire College, Victoria University; and Herbert Perkin, M.I.M.E., Certificated Colliery Manager, Assistant Lecturer in the Mining Department of the Yorkshire College, Victoria University. Second Edition, Revised and Enlarged. Medium 8vo, cloth, 300 pp., with about 170 Illustrations . . . . *Net* **12/0**

Introductory—Dynamic Electricity—Driving of the Dynamo—The Steam Turbine—Distribution of Electrical Energy—Starting and Stopping Electrical Generators and Motors—Electric Cables—Central Electrical Plants—Electricity Applied to Pumping and Hauling—Electricity Applied to Coal Cutting—Typical Electric Plants Recently Erected—Electric Lighting by Arc and Glow Lamps—Miscellaneous Applications of Electricity—Electricity as Compared with Other Modes of Transmitting Power—Dangers of Electricity.

"The book is a good attempt to meet a growing want, and is well worthy of a place in the mining engineer's library."—*The Electrician.*

ELECTRICITY. A STUDENT'S TEXT-BOOK. By H. M. Noad, F.R.S. 650 pp., with 470 Illustrations. Crown 8vo . . **9/0**

ELECTRICITY, POWER TRANSMITTED BY, AND APPLIED BY THE ELECTRIC MOTOR, including Electric Railway Construction. By Philip Atkinson, A.M., Ph.D., author of "Elements of Static Electricity." Fourth Edition, Enlarged, Crown 8vo, cloth, 224 pp., with over 90 illustrations . . . . *Net* **9/0**

ELECTRO-PLATING AND ELECTRO-REFINING OF METALS. Being a new edition of Alexander Watt's "Electro-Deposition." Revised and Largely Re-written by Arnold Philip, Assoc. R.S.M., B.Sc., A.I.E.E., F.I.C., Principal Assistant to the Admiralty Chemist, formerly Chief Chemist to the Engineering Departments of the India Office, and sometime Assistant Professor of Electrical Engineering and Applied Physics at the Heriot-Watt College, Edinburgh. 700 pp., with numerous Illustrations, Large Crown 8vo, cloth . . . . *Net* **12/6**

ENGINE-DRIVING LIFE. Stirring Adventures and Incidents in the Lives of Locomotive Engine-Drivers. By Michael Reynolds. Third Edition. Crown 8vo, cloth **1/6**

ENGINEERING DRAWING. A WORKMAN'S MANUAL. By John Maxton, Instructor in Engineering Drawing, Royal Naval College, Greenwich. Eighth Edition. 300 Plates and Diagrams. Crown 8vo, cloth **3/6**

"A copy of it should be kept for reference in every drawing office."—*Engineering.*

ENGINEERING ESTIMATES, COSTS, AND ACCOUNTS. A Guide to Commercial Engineering. With numerous examples of Estimates and Costs of Millwright Work, Miscellaneous Productions, Steam Engines and Steam Boilers; and a Section on the Preparation of Costs Accounts. By A General Manager. Second Edition. 8vo, cloth. . . . **12/0**

"The information is given in a plain, straightforward manner, and bears throughout evidence of the intimate practical acquaintance of the author with every phase of commercial engineering"
—*Mechanical World.*

ENGINEERING PROGRESS (1863-6). By WM. HUMBER, A.M.Inst.C.E. Complete in Four Vols. Containing 148 Double Plates, with Portraits and Copious Descriptive Letterpress. Impl. 4to, half-morocco. Price, complete, **£12 12s.**; or each Volume sold separately at **£3 3s.** per Volume. *Descriptive List of Contents on application.*

ENGINEERING STANDARDS COMMITTEE'S PUBLICATIONS. See pages 31 and 32.

ENGINEER'S AND MILLWRIGHT'S ASSISTANT. A Collection of Useful Tables, Rules, and Data. By WILLIAM TEMPLETON. Eighth Edition, with Additions. 18mo, cloth **2/6**

"A deservedly popular work. It should be in the 'drawer' of every mechanic."—*English Mechanic.*

ENGINEER'S HANDBOOK. A Practical Treatise on Modern Engines and Boilers, Marine, Locomotive, and Stationary. And containing a large collection of Rules and Practical Data relating to Recent Practice in Designing and Constructing all kinds of Engines, Boilers, and other Engineering work. The whole constituting a comprehensive Key to the Board of Trade and other Examinations for Certificates of Competency in Modern Mechanical Engineering. By WALTER S. HUTTON, Civil and Mechanical Engineer, Author of "The Works' Manager's Handbook for Engineers," &c. With upwards of 420 Illustrations. Sixth Edition, Revised and Enlarged. Medium 8vo, nearly 560 pp., strongly bound **18/0**

"A mass of information set down in simple language, and in such a form that it can be easily referred to at any time. The matter is uniformly good and well chosen, and is greatly elucidated by the illustrations. The book will find its way on to most engineers' shelves, where it will rank as one of the most useful books of reference."—*Practical Engineer.*

"Full of useful information, and should be found on the office shelf of all practical engineers."—*English Mechanic.*

ENGINEER'S, MECHANIC'S, ARCHITECT'S, BUILDER'S, ETC. TABLES AND MEMORANDA. Selected and Arranged by FRANCIS SMITH. Seventh Edition, Revised, including ELECTRICAL TABLES, FORMULÆ, and MEMORANDA. Waistcoat-pocket size, limp leather **1/6**

"The best example we have ever seen of 270 pages of useful matter packed into the dimensions of a card-case."—*Building News.*

ENGINEER'S YEAR-BOOK FOR 1907. Comprising Formulæ, Rules, Tables, Data and Memoranda in Civil, Mechanical, Electrical, Marine and Mine Engineering. By H. R. KEMPE, M.Inst.C.E., Principal Staff Engineer, Engineer-in-Chief's Office, General Post Office, London, Author of "A Handbook of Electrical Testing," "The Electrical Engineer's Pocket-Book," &c. With 1,000 Illustrations, specially Engraved for the work. Crown 8vo, 950 pp., leather. [*Just Published.* **8/0**

"Kempe's Year-Book really requires no commendation. Its sphere of usefulness is widely known, and it is used by engineers the world over."—*The Engineer.*

"The volume is distinctly in advance of most similar publications in this country."—*Engineering.*

ENGINEMAN'S POCKET COMPANION, and Practical Educator for Enginemen, Boiler Attendants, and Mechanics. By MICHAEL REYNOLDS. With 45 Illustrations and numerous Diagrams. Fifth Edition. Royal 18mo, strongly bound for pocket wear **3/6**

EXCAVATION (EARTH AND ROCK). A Practical Treatise, by CHARLES PRELINI, C.E. 365 pp., with Tables, many Diagrams and Engravings. Royal 8vo, cloth. . . . . . . *Net* **16/0**

FACTORY ACCOUNTS: their PRINCIPLES & PRACTICE. A Handbook for Accountants and Manufacturers. By E. GARCKE and J. M. FELLS. Crown 8vo, cloth **7/6**

FIRES, FIRE-ENGINES, AND FIRE BRIGADES. With a History of Fire-Engines, their Construction, Use, and Management. Hints on Fire-Brigades, &c. By C. F. T. YOUNG, C.E. 8vo, cloth, **£1 4s.**

FOUNDATIONS AND CONCRETE WORKS. With Practical Remarks on Footings, Planking, Sand and Concrete, Béton, Pile-driving, Caissons, and Cofferdams. By E. DOBSON. Crown 8vo. . . . **1/6**

FUEL, ITS COMBUSTION AND ECONOMY. Consisting of an Abridgment of "A Treatise on the Combustion of Coal and the Prevention of Smoke." By C. W. WILLIAMS, A.Inst.C.E. With extensive Additions by D. KINNEAR CLARK, M.Inst.C.E. Fourth Edition. Crown 8vo, cloth **3/6**

FUELS: SOLID, LIQUID, AND GASEOUS. Their Analysis and Valuation. For the use of Chemists and Engineers. By H. J. PHILLIPS' F.C.S., formerly Analytical and Consulting Chemist to the Great Eastern Railway. Fourth Edition. Crown 8vo, cloth **2/0**

GAS AND OIL ENGINE MANAGEMENT. A Practical Guide for Users and Attendants, being Notes on Selection, Construction, and Management. By M. POWIS BALE, M.Inst.C.E., M.I Mech.E. Author of "Woodworking Machinery," &c. Second Edition, with an additional Chapter on Gas Producers. Crown 8vo, cloth . . . . . *Net* **3/6**

SELECTING AND FIXING A GAS ENGINE—PRINCIPLES OF WORKING, ETC., FAILURES AND DEFECTS—VALVES, IGNITION, PISTON RINGS, ETC.—OIL ENGINES—GAS PRODUCERS—RULES, TABLES, ETC.

GAS ENGINEER'S POCKET-BOOK. Comprising Tables, Notes and Memoranda relating to the Manufacture, Distribution, and Use of Coal gas and the Construction of Gas Works. By H. O'CONNOR, A.M.Inst. C.E. Third Edition, Revised. Crown 8vo, leather.
[*Just Published. Net* **10/6**
"The book contains a vast amount of information "—*Gas World*.

THE GAS-ENGINE HANDBOOK. A Manual of Useful Information for the Designer and the Engineer. By E. W. ROBERTS, M.E. With Forty Full-page Engravings. Small Fcap. 8vo, leather . . *Net* **8/6**

GAS-ENGINES AND PRODUCER-GAS PLANTS. A Treatise setting forth the Principles of Gas Engines and Producer Design, the Selection and Installation of an Engine, Conditions of Perfect Operation, Producer-Gas Engines and their Possibilities, the Care of Gas Engines and Producer-Gas Plants, with a Chapter on Volatile Hydrocarbon and Oil Engines. By R. E. MATHOT, M.E. Translated from the French. With a Preface by DUGALD CLERK, M.Inst.C.E., F.C.S. Medium 8vo, cloth, 310 pages, with about 150 Illustrations . . . . . *Net* **12/0**

MOTIVE POWER AND COST OF INSTALLATION—SELECTION OF AN ENGINE—INSTALLATION OF AN ENGINE—FOUNDATION AND EXHAUST—WATER CIRCULATION—LUBRICATION—CONDITIONS OF PERFECT OPERATION—HOW TO START AN ENGINE—PRECAUTIONS—PERTURBATIONS IN THE OPERATIONS OF ENGINES—PRODUCER-GAS ENGINES—PRODUCER-GAS—PRESSURE GAS-PRODUCERS—SUCTION GAS-PRODUCERS—OIL AND VOLATILE HYDROCARBON ENGINES—THE SELECTION OF AN ENGINE.

GAS ENGINES. With Appendix describing a Recent Engine with Tube Igniter. By T. M. GOODEVE, M.A. Crown 8vo, cloth . **2/6**

GAS MANUFACTURE, CHEMISTRY OF. A Practical Manual for the Use of Gas Engineers, Gas Managers and Students. By HAROLD M. ROYLE, F.C.S., Chief Chemical Assistant at the Becton Gas Works . . . . . . [*Nearly ready.* Price about **12/6** *net.*

GAS WORKS. Their Construction and arrangement, and the Manufacture and Distribution of Coal Gas. By S. HUGHES, C.E. Ninth Edition. Revised, with Notices of Recent Improvements by HENRY O'CONNOR, A.M.Inst.C.E. Crown 8vo, cloth **6/-**

GEOMETRY. For the Architect, Engineer, and Mechanic, By E. W. TARN, M.A., Architect. 8vo, cloth **9/0**

GEOMETRY FOR TECHNICAL STUDENTS. An Introduction to Pure and Applied Geometry and the Mensuration of Surfaces and Solids, including Problems in Plane Geometry useful in Drawing. By E. H. SPRAGUE, A.M.Inst.C.E. Crown 8vo, cloth. . . . . *Net* **1/0**

GEOMETRY OF COMPASSES; or Problems Resolved by the mere Description of Circles and the Use of Coloured Diagrams and Symbols. By OLIVER BYRNE. Coloured Plates. Crown 8vo, cloth . **3/6**

HEAT, EXPANSION OF STRUCTURES BY. By JOHN KEILY, C.E. Crown 8vo, cloth **3/6**

HOISTING MACHINERY. An Elementary Treatise on. Including the Elements of Crane Construction and Descriptions of the Various Types of Cranes in Use. By JOSEPH HORNER, A.M.I.M.E., Author of "Pattern-Making," and other Works. Crown 8vo, cloth, with 215 Illustrations, including Folding Plates . . . . . . *Net* **7/6**

HYDRAULIC MANUAL. Consisting of Working Tables and Explanatory Text. Intended as a Guide in Hydraulic Calculations and Field Operations. By LOWIS D'A. JACKSON. Fourth Edition, Enlarged. Large crown 8vo, cloth **16/0**

HYDRAULIC POWER ENGINEERING. A Practical Manual on the Concentration and Transmission of Power by Hydraulic Machinery. By G. CROYDON MARKS, A.M.Inst.C.E. Second Edition, Enlarged, with about 240 Illustrations. 8vo, cloth . . . . *Net* **10/6**

SUMMARY OF CONTENTS:—PRINCIPLES OF HYDRAULICS.—THE FLOW OF WATER.—HYDRAULIC PRESSURES.—MATERIAL.—TEST LOAD.—PACKING FOR SLIDING SURFACES.—PIPE JOINTS.—CONTROLLING VALVES.—PLATFORM LIFTS.—WORKSHOP AND FOUNDRY CRANES.—WAREHOUSE AND DOCK CRANES.—HYDRAULIC ACCUMULATORS.—PRESSES FOR BALING AND OTHER PURPOSES.—SHEET METAL WORKING AND FORGING MACHINERY.—HYDRAULIC RIVETERS.—HAND AND POWER PUMPS.—STEAM PUMPS.—TURBINES.—IMPULSE TURBINES —REACTION TURBINES.—DESIGN OF TURBINES IN DETAIL.—WATER WHEELS.—HYDRAULIC ENGINES.—RECENT ACHIEVEMENTS.—PRESSURE OF WATER.—ACTION OF PUMPS, &c.

"Can be unhesitatingly recommended as a useful and up-to-date manual on hydraulic transmission and utilisation of power."—*Mechanical World*.

HYDRAULIC TABLES, CO-EFFICIENTS, & FORMULÆ. For Finding the Discharge of Water from Orifices, Notches, Weirs, Pipes, and Rivers. With New Formulæ, Tables, and General Information on Rain-fall, Catchment-Basins, Drainage, Sewerage, Water Supply for Towns and Mill Power. By JOHN NEVILLE, C.E., M.R.I.A. Third Edition, revised, with additions. Numerous Illustrations. Crown 8vo, cloth . . . **14/0**

IRON AND METAL TRADES' COMPANION. For Expeditiously Ascertaining the Value of any Goods bought or sold by Weight, from 1s. per cwt. to 112s. per cwt., and from one farthing per pound to one shilling per pound. By THOMAS DOWNIE. Strongly bound in leather, 396 pp. **9/0**

IRON AND STEEL. A Work for the Forge, Foundry, Factory, and Office. Containing ready, useful, and trustworthy Information for Ironmasters and their Stock-takers; Managers of Bar, Rail, Plate, and Sheet Rolling Mills; Iron and Metal Founders; Iron Ship and Bridge Builders; Mechanical, Mining, and Consulting Engineers; Architects, Contractors, Builders, &c. By CHARLES HOARE, Author of "The Slide Rule," &c. Ninth Edition. 32mo, leather **6/0**

IRON AND STEEL CONSTRUCTIONAL WORK, as applied to Public, Private, and Domestic Buildings. By FRANCIS CAMPIN, C.E. Crown 8vo, cloth **3/6**

IRON & STEEL GIRDERS. A Graphic Table for facilitating the Computation of the Weights of Wrought Iron and Steel Girders, &c., for Parliamentary and other Estimates. By J. H. WATSON BUCK, M.Inst.C.E. On a Sheet **2/6**

IRON-PLATE WEIGHT TABLES. For Iron Shipbuilders, Engineers, and Iron Merchants. Containing the Calculated Weights of upwards of 150,000 different sizes of Iron Plates from 1 foot by 6 in. by ¼ in. to 10 feet by 5 feet by 1 in. Worked out on the Basis of 40 lbs. to the square foot of Iron of 1 inch in thickness. By H. BURLINSON and W. H. SIMPSON. 4to, half-bound . . . . **£1 5s.**

IRRIGATON (PIONEER). A Manual of Information for Farmers in the Colonies. By E. O. MAWSON, M.Inst.C.E., Executive Engineer, Public Works Department, Bombay. With Chapters on Light Railways by E. R. CALTHROP, M.Inst.C.E., M.I.M.E. With Plates and Diagrams. Demy 8vo, cloth **10/6**
VALUE OF IRRIGATION, AND SOURCES OF WATER SUPPLY—DAMS AND WEIRS—CANALS—UNDERGROUND WATER—METHODS OF IRRIGATION—SEWAGE IRRIGATION—IMPERIAL AUTOMATIC SLUICE GATES—THE CULTIVATION OF IRRIGATED CROPS, VEGETABLES, AND FRUIT TREES—LIGHT RAILWAYS FOR HEAVY TRAFFIC—USEFUL MEMORANDA AND DATA.

LATHE PRACTICE. A complete and Practical Work on the Modern American Lathe. By OSCAR E. PERRIGO, M.E., Author of "Modern Machine Shop Construction, Equipment, and Management," etc. Medium 8vo, 424 pp., 315 illustrations. Cloth.
[*Just Published.* Net **12/0**
HISTORY OF THE LATHE UP TO THE INTRODUCTION OF SCREW THREADS—ITS DEVELOPMENT SINCE THE INTRODUCTION OF SCREW THREADS—CLASSIFICATION OF LATHES—LATHE DESIGN: THE BED AND ITS SUPPORTS—THE HEAD-STOCK CASTING, THE SPINDLE, AND SPINDLE-CONE—THE SPINDLE BEARINGS, THE BACK GEARS, AND THE TRIPLE-GEAR MECHANISM—THE TAIL STOCK, THE CARRIAGE, THE APRON, ETC.—TURNING RESTS, SUPPORTING RESTS, SHAFT STRAIGHTENERS, ETC.—LATHE ATTACHMENTS—RAPID CHANGE GEAR MECHANISM—LATHE TOOLS, HIGH-SPEED STEEL, SPEEDS AND FEEDS, POWER FOR CUTTING TOOLS, ETC.—TESTING A LATHE—LATHE WORK—ENGINE LATHES—HEAVY LATHES—HIGH-SPEED LATHES—SPECIAL LATHES—REGULAR TURRET LATHES—SPECIAL TURRET LATHES—ELECTRICALLY-DRIVEN LATHES.

LATHE-WORK. A Practical Treatise on the Tools, Appliances, and Processes employed in the Art of Turning. By PAUL N. HASLUCK. Eighth Edition. Crown 8vo, cloth **5/0**
"We can safely recommend the work to young engineers. To the amateur it will simply be invaluable.—*Engineer.*

LAW FOR ENGINEERS AND MANUFACTURERS. See EVERY MAN'S OWN LAWYER. A Handybook of the Principles of Law and Equity. By a Barrister. Forty-fourth (1907) Edition, Revised and Enlarged, including Abstracts of the Legislature of 1906 of especial interest to Engineering Firms and Manufacturers, such as the Workmen's Compensation Act, 1906; the Prevention of Corruption Act, 1906; the Trades Disputes Act, 1906; the Merchant Shipping Act, 1906; the Marine Insurance Act, 1906, and many other recent Acts. Large crown 8vo, cloth, 838 pages.
Just published. Net **6/8**
"No Englishman ought to be without this book."—*Engineer.*
"Ought to be in every business establishment and in all libraries."—*Sheffield Post.*
"It is a complete code of English Law written in plain language, which all can understand. . . Should be in the hands of every business man, and all who wish to abolish lawyers' bills."—*Weekly Times.*
"A useful and concise epitome of the law, compiled with considerable care."—*Law Magazine.*

LEVELLING, PRINCIPLES AND PRACTICE OF. Showing its Application to Purposes of Railway and Civil Engineering in the Construction of Roads; with Mr. TELFORD'S Rules for the same. By FREDERICK W. SIMMS, M.Inst.C.E. Ninth Edition, with LAW'S Practical Examples for Setting-out Railway Curves, and TRAUTWINE'S Field Practice of Laying-out Circular Curves. With 7 Plates and numerous Woodcuts, 8vo **8/6**
"The publishers have rendered a substantial service to the profession, especially to the younger members, by bringing out the present edition of Mr. Simms's useful work."—*Engineering.*

LOCOMOTIVE ENGINE. The Autobiography of an old Locomotive Engine. By ROBERT WEATHERBURN, M.I.M.E. With Illustrations and Portraits of GEORGE and ROBERT STEPHENSON. Crown 8vo, cloth.
Net **2/6**

LOCOMOTIVE ENGINE DEVELOPMENT. A Popular Treatise on the Gradual Improvements made in Railway Engines between 1803 and 1903. By CLEMENT E. STRETTON, C.E. Sixth Edition, Revised and Enlarged. Crown 8vo, cloth . . . . . *Net* **4/6**

LOCOMOTIVE ENGINE DRIVING. A Practical Manual for Engineers in Charge of Locomotive Engines. By MICHAEL REYNOLDS, M.S.E. Twelfth Edition. Crown 8vo, cloth, **3/6**; cloth boards **4/6**

LOCOMOTIVE ENGINES. A Rudimentary Treatise on. By G. D. DEMPSEY, C.E. With large Additions treating of the Modern Locomotive, by D. K. CLARK, M.Inst.C.E. With Illustrations. Crown 8vo, cloth **3/0**

LOCOMOTIVE (MODEL) ENGINEER, Fireman and Engine-Boy. Comprising a Historical Notice of the Pioneer Locomotive Engines and their Inventors. By MICHAEL REYNOLDS. New Edition, with Revised Appendix. Crown 8vo, cloth, **3/6**; cloth boards **4/6**

MACHINERY, DETAILS OF. Comprising Instructions for the Execution of various Works in Iron in the Fitting Shop, Foundry, and Boiler Yard. By FRANCIS CAMPIN, C.E. Crown 8vo, cloth . . **3/0**

MACHINE SHOP TOOLS. A Practical Treatise describing in every detail the Construction, Operation and Manipulation of both Hand and Machine Tools; being a work of Practical Instruction in all Classes of Modern Machine Shop Practice, including Chapters on Filing, Fitting and Scraping Surfaces; on Drills, Reamers, Taps and Dies; the Lathe and its Tools; Planers, Shapers and their Tools; Milling Machines and Cutters; Gear Cutters and Gear Cutting; Drilling Machines and Drill Work; Grinding Machines and their Work; Hardening and Tempering, Gearing, Belting, and Transmission Machinery; Useful Data and Tables. By WILLIAM H. VAN DERVOORT, M.E. Fourth Edition. Illustrated by 673 Engravings. Medium 8vo, cloth . . . . *Net* **21/0**

MARINE ENGINEERING. An Elementary Manual for Young Marine Engineers and Apprentices. By J. S. BREWER. Crown 8vo, cloth. **1/6**

"A useful introduction to the more elaborate text-books."—*Scotsman.*

MARINE ENGINEER'S GUIDE to Board of Trade Examinations for Certificates of Competency. Containing all Latest Questions to Date, with Simple, Clear, and Correct Solutions; 302 Elementary Questions with Illustrated Answers, and Verbal Questions and Answers; complete Set of Drawings with Statements completed. By A. C. WANNAN, C.E., Consulting Engineer, and E. W. I. WANNAN, M.I.M.E., Certificated First Class Marine Engineer. With numerous Engravings. Fourth Edition, Enlarged. 500 pages. Large crown 8vo, cloth . . . . . *Net* **10/6**

"The book is clearly and plainly written and avoids unnecessary explanations and formulas, and we consider it a valuable book for students of marine engineering."—*Nautical Magazine.*

MARINE ENGINEER'S POCKET-BOOK. Containing latest Board of Trade Rules and Data for Marine Engineers. By A. C. WANNAN. Fourth Edition, Revised, Enlarged, and Brought up to Date. Square 18mo, with thumb Index, leather **5/0**

MARINE ENGINES AND BOILERS. Their Design and Construction. A Handbook for the Use of Students, Engineers, and Naval Constructors. Based on the Work "Berechnung und Konstruktion der Schiffsmaschinen und Kessel," by Dr. G. BAUER, Engineer-in-Chief of the Vulcan Shipbuilding Yard, Stettin. Translated from the Second German Edition by E. M. DONKIN, and S. BRYAN DONKIN, A.M.I.C.E. Edited by LESLIE S. ROBERTSON, Secretary to the Engineering Standards Committee, M.I.C.E., M.I.M.E., M.I.N.A., &c. With numerous Illustrations and Tables. Medium 8vo, cloth . . . . . . **25/- *Net.***

SUMMARY OF CONTENTS:—PART I.—MAIN ENGINES.—DETERMINATION OF CYLINDER DIMENSIONS.—THE UTILISATION OF STEAM IN THE ENGINE.—STROKE OF PISTON.—NUMBER OF REVOLUTIONS.—TURNING MOMENT.—BALANCING OF THE MOVING PARTS.—ARRANGEMENT OF MAIN ENGINES.—DETAILS OF MAIN ENGINES.—THE CYLINDER.—VALVES.—VARIOUS KINDS OF VALVE GEAR.—PISTON RODS.—PISTONS.—CONNECTING ROD AND CROSSHEAD.—VALVE GEAR RODS.—BED PLATES.—ENGINE COLUMNS,—REVERSING AND TURNING GEAR. PART II.—PUMPS.—AIR, CIRCULATING FEED, AND AUXILIARY PUMPS. PART III.—SHAFTING, RESISTANCE OF SHIPS, PROPELLERS.—THRUST SHAFT AND THRUST BLOCK.—TUNNEL SHAFTS AND PLUMMER BLOCKS.—SHAFT COUPLINGS.—STERN TUBE.—THE SCREW PROPELLER.—CONSTRUCTION OF THE SCREW. PART IV.—PIPES AND CONNECTIONS.—GENERAL REMARKS, FLANGES, VALVES, &c.—UNDER WATER FITTINGS.—MAIN STEAM, AUXILIARY STEAM, AND EXHAUST PIPING.—FEED WATER, BILGE, BALLAST AND CIRCULATING PIPES. PART V.—STEAM BOILERS.—FIRING AND THE GENERATION OF STEAM.—CYLINDRICAL BOILERS.—LOCOMOTIVE BOILERS.—WATER-TUBE BOILERS.—SMALL TUBE WATER-TUBE BOILERS.—SMOKE BOX.—FUNNEL AND BOILER LAGGING.—FORCED DRAUGHT.—BOILER FITTINGS AND MOUNTINGS. PART VI.—MEASURING INSTRUMENTS. PART VII.—VARIOUS DETAILS.—BOLTS, NUTS, SCREW THREADS, &c.—PLATFORMS, GRATINGS, LADDERS.—FOUNDATIONS.—SEATINGS.—LUBRICATION.—VENTILATION OF ENGINE ROOMS.—RULES FOR SPARE GEAR. PART VIII.—ADDITIONAL TABLES.

"This handsome volume contains a comprehensive account of the design and construction of modern marine engines and boilers. Its arrangement is excellent, and the numerous illustrations represent recent practice for all classes of warships and vessels of the mercantile marine. His position as Engineer-in-Chief of the great Vulcan Works at Stettin gave the author special facilities for selecting illustrations from the practice of that firm, which has built many of the swiftest types of steamships for both war and commerce. Other German firms and the German Admiralty have been equally generous in contributing information, while a large proportion of the illustrations is drawn from English technical journals and the proceedings of our engineering societies. American practice is also represented. The compilation has been laborious, no doubt, but it constitutes a valuable book of reference and a treasury of information. The English editor and his assistants have done their work well, both in translation and in the conversion of metric to English measures."
—*The Times.*

MARINE ENGINES AND STEAM VESSELS. By R. MURRAY, C.E. Eighth Edition, thoroughly Revised, with Additions by the Author and by GEORGE CARLISLE, C.E. Crown 8vo, cloth . . **4/6**

MASONRY DAMS FROM INCEPTION TO COMPLETION. Including numerous Formulæ, Forms of Specification and Tender, Pocket Diagram of Forces, &c. For the use of Civil and Mining Engineers. By C. F. COURTNEY, M.Inst.C.E. 8vo, cloth **9/0**

MASTING, MAST-MAKING, AND RIGGING OF SHIPS. Also Tables of Spars, Rigging, Blocks; Chain, Wire, and Hemp Ropes, &c., relative to every class of vessels. By R. KIPPING. Crown 8vo, cloth **2/0**

MATERIALS AND CONSTRUCTION. A Theoretical and Practical Treatise on the Strains, Designing, and Erection of Works of Construction. By F. CAMPIN. Crown 8vo, cloth **3/0**

MATERIALS, A TREATISE ON THE STRENGTH OF. By P. BARLOW, F.R.S., P. W. BARLOW, F.R.S., and W. H. BARLOW, F.R.S. Edited by WM. HUMBER, A.M.Inst.C.E. 8vo, cloth **18/0**
'The standard treatise on that particular subject."—*Engineer.*

MATHEMATICAL TABLES. For Trigonometrical, Astronomical, and Nautical Calculations; to which is prefixed a Treatise on Logarithms, by H. LAW, C.E. With Tables for Navigation and Nautical Astronomy. By Prof. J. R. YOUNG. Crown 8vo, cloth . . . **4/0**

MECHANICAL ENGINEERING. Comprising Metallurgy, Moulding, Casting, Forging, Tools, Workshop Machinery, Mechanical Manipulation, Manufacture of the Steam Engine, &c. By FRANCIS CAMPIN, C.E. Third Edition. Crown 8vo, cloth **2/6**

MECHANICAL ENGINEERING TERMS. (Lockwood's Dictionary). Embracing terms current in the Drawing Office, Pattern Shop, Foundry, Fitting, Turning, Smiths', and Boiler Shops, &c. Comprising upwards of 6,000 Definitions. Edited by J. G. HORNER, A.M.I.M.E. Third Edition, Revised, with Additions. Crown 8vo, cloth . . . *Net* **7/6**

THE MECHANICAL ENGINEER'S COMPANION. Areas, Circumferences, Decimal Equivalents, in inches and feet, millimetres, squares, cubes, roots, &c.; Strength of Bolts, Weight of Iron, &c.; Weights, Measures, and other Data. Also Practical Rules for Engine Proportions. By R. EDWARDS, M.Inst.C.E. Fcap. 8vo, cloth. **3/6**

MECHANICAL ENGINEER'S POCKET-BOOK. Comprising Tables, Formulæ, Rules, and Data: A Handy Book of Reference for Daily Use in Engineering Practice. By D. KINNEAR CLARK, M.Inst.C.E., Sixth Edition, thoroughly Revised and Enlarged. By H. H. P. POWLES, A.M.Inst.C.E., M.I.M.E. Small 8vo, 700 pp., Leather
[*Just published. Net* **6/0**
MATHEMATICAL TABLES.—MEASUREMENT OF SURFACES AND SOLIDS.—ENGLISH WEIGHTS AND MEASURES.—FRENCH METRIC WEIGHTS AND MEASURES.—FOREIGN WEIGHTS AND MEASURES.—MONEYS.—SPECIFIC GRAVITY, WEIGHT, AND VOLUME.—MANUFACTURED METALS.—STEEL PIPES.—BOLTS AND NUTS.—SUNDRY ARTICLES IN WROUGHT AND CAST IRON, COPPER, BRASS, LEAD, TIN, ZINC.—STRENGTH OF MATERIALS.—STRENGTH OF TIMBER.—STRENGTH OF CAST IRON.—STRENGTH OF WROUGHT IRON.—STRENGTH OF STEEL.—TENSILE STRENGTH OF COPPER, LEAD, &c.—RESISTANCE OF STONES AND OTHER BUILDING MATERIALS.—RIVETED JOINTS IN BOILER PLATES.—BOILER SHELLS.—WIRE ROPES AND HEMP ROPES.—CHAINS AND CHAIN CABLES.—FRAMING.—HARDNESS OF METALS, ALLOYS, AND STONES.—LABOUR OF ANIMALS.—MECHANICAL PRINCIPLES.—GRAVITY AND FALL OF BODIES.—ACCELERATING AND RETARDING FORCES.—MILL GEARING, SHAFTING, &c.—TRANSMISSION OF MOTIVE POWER.—HEAT.—COMBUSTION: FUELS.—WARMING, VENTILATION, COOKING STOVES.—STEAM.—STEAM ENGINES AND BOILERS.—RAILWAYS.—TRAMWAYS.—STEAM SHIPS.—PUMPING STEAM ENGINES AND PUMPS.—COAL GAS—GAS ENGINES, &c.—AIR IN MOTION—COMPRESSED AIR.—HOT AIR ENGINES.—WATER POWER.—SPEED OF CUTTING TOOLS.—COLOURS.—ELECTRICAL ENGINEERING.
"It would be found difficult to compress more matter within a similar compass, or produce a book of 700 pages which should be more compact or convenient for pocket reference. . . . Will be appreciated by mechanical engineers of all classes."—*Practical Engineer.*

MECHANICAL ENGINEER'S REFERENCE BOOK. For Machine and Boiler Construction. By NELSON FOLEY, M.I.N.A. New Edition, Revised throughout and much Enlarged. To be issued in parts.
[*In Preparation.*
"Mr. Foley is well fitted to compile such a work. The diagrams are a great feature of the work. It may be stated that Mr. Foley has produced a volume which will undoubtedly fulfil the desire of the author and become indispensable to all mechanical engineers."—*Marine Engineer.*
"We have carefully examined this work, and pronounce it a most excellent reference book for the use of marine engineers."—*Journal of American Society of Naval Engineers.*

THE MECHANICAL HANDLING OF MATERIAL. A Treatise on the Handling of Material such as Coal, Ore, Timber, &c., by Automatic or Semi-Automatic Machinery, together with the Various Accessories used in the Manipulation of such Plant, and Dealing fully with the Handling, Storing, and Warehousing of Grain. By G. F. ZIMMER, A.M.Inst.C.E. 528 pages Royal 8vo, cloth, with 550 Illustrations (including Folding Plates) specially prepared for the Work . . . *Net* **25/0**

"It is an essentially practical work written by a practical man, who is not only thoroughly acquainted with his subject theoretically, but who also has the knowledge that can only be obtained by actual experience in working and planning installations for the mechanical handling of raw material."—*The Times.*

MECHANICS. Being a concise Exposition of the General Principles of Mechanical Science and their Applications. By C. TOMLINSON, F.R.S. Crown 8vo, cloth **1/6**

MECHANICS: CONDENSED. A Selection of Formulæ, Rules, Tables, and Data for the Use of Engineering Students, &c. By W. G. C. HUGHES, A.M.I.C.E. Crown 8vo, cloth **2/6**

MECHANICS OF AIR MACHINERY. By Dr. J. WIESBACH and PROF. G. HERRMANN. Authorized Translation with an Appendix on American Practice by A. TROWBRIDGE, Ph.B., Adjunct Professor of Mechanical Engineering, Columbia University. Royal 8vo, cloth. . . *Net* **18/0**

MECHANICS' WORKSHOP COMPANION. Comprising a great variety of the most useful Rules and Formulæ in Mechanical Science, with numerous Tables of Practical Data and Calculated Results for Facilitating Mechanical Operations. By WILLIAM TEMPLETON, Author of "The Engineer's Practical Assistant," &c., &c. Eighteenth Edition, Revised, Modernised, and considerably Enlarged by W. S. HUTTON, C.E., Author of "The Works' Manager's Handbook," &c. Fcap. 8vo, nearly 500 pp., with 8 Plates and upwards of 250 Diagrams, leather **6/0**

"This well-known and largely-used book contains information, brought up to date, of the sort so useful to the foreman and draughtsman. So much fresh information has been introduced as to constitute it practically a new book."—*Mechanical World.*

MECHANISM AND MACHINE TOOLS. By T. BAKER, C.E. With Remarks on Tools and Machinery by J. NASMYTH, C.E. Crown 8vo, cloth **2/6**

MENSURATION AND MEASURING. With the Mensuration and Levelling of Land for the purposes of Modern Engineering. By T. BAKER, C.E. New Edition by E. NUGENT, C.E. Crown 8vo, cloth **1/6**

METAL-TURNING. A Practical Handbook for Engineers, Technical Students, and Amateurs. By JOSEPH HORNER, A.M.I.Mech.E., Author of "Pattern Making," &c. Large Crown 8vo, cloth, with 488 Illustrations. [*Just Published.* *Net* **9/0**

SUMMARY OF CONTENTS:—INTRODUCTION.—RELATIONS OF TURNERY AND MACHINE SHOP.—SEC. I. THE LATHE, ITS WORK, AND TOOLS.—FORMS AND FUNCTIONS OF TOOLS —REMARKS ON TURNING IN GENERAL.—SEC. II. TURNING BETWEEN CENTRES.—CENTRING AND DRIVING.—USE OF STEADIES.—EXAMPLES OF TURNING INVOLVING LINING-OUT FOR CENTRES.—MANDREL WORK.—SEC. III. WORK SUPPORTED AT ONE END.—FACE PLATE TURNING.—ANGLE PLATE TURNING.—INDEPENDENT JAW CHUCKS.—CONCENTRIC, UNIVERSAL, TOGGLE, AND APPLIED CHUCKS.—SEC. IV. INTERNAL WORK.—DRILLING, BORING, AND ALLIED OPERATIONS.—SEC. V. SCREW CUTTINGS AND TURRET WORK. —SEC. VI MISCELLANEOUS.—SPECIAL WORK.—MEASUREMENT, GRINDING.—TOOL HOLDERS—SPEED AND FEEDS, TOOL STEELS.—STEEL MAKERS' INSTRUCTIONS.

METRIC TABLES. In which the British Standard Measures and Weights are compared with those of the Metric System at present in Use on the Continent. By C. H. DOWLING, C.E. 8vo, cloth . . . **10/6**

MILLING MACHINES; their Design, Construction, and Working. A Handbook for Practical Men and Engineering Students. By JOSEPH HORNER, A.M.I.Mech.E., Author of "Pattern Making," &c. With 269 Illustrations. Medium 8vo, cloth . . . . . *Net* **12/6**
LEADING ELEMENTS OF MILLING MACHINE DESIGN AND CONSTRUCTION—PLAIN AND UNIVERSAL MACHINES — ATTACHMENTS AND BRACINGS — VERTICAL SPINDLES MACHINES—PLANO-MILLERS OR SLABBING MACHINES—SPECIAL MACHINES—CUTTERS—MILLING OPERATIONS—INDEXING, SPIRAL WORK, AND WORM, SPUR, AND BEVEL GEARS, ETC.—SPUR AND BEVEL GEARS—FEEDS AND SPEEDS.

MOTOR CARS FOR COMMON ROADS. By A. J. WALLIS-TAYLER, A.M.Inst.C.E. 212 pp., with 76 Illustrations. Crown 8vo, cloth **4/6**

MOTOR VEHICLES FOR BUSINESS PURPOSES. A Practical Handbook for those interested in the Transport of Passengers and Goods. By A. J. WALLIS-TAYLER, A.M.Inst.C.E. With 134 Illustrations. Demy 8vo, cloth. . . . . *Net* **9/0**

NAVAL ARCHITECT'S AND SHIPBUILDER'S POCKET BOOK. Of Formulæ, Rules, and Tables, and Marine Engineer's and Surveyor's Handy Book of Reference. By CLEMENT MACKROW, M.I.N.A. Ninth Edition, Fcap., leather . . . . *Net* **12/6**
SIGNS AND SYMBOLS, DECIMAL FRACTIONS,—TRIGONOMETRY.—PRACTICAL GEOMETRY.—MENSURATION.—CENTRES AND MOMENTS OF FIGURES.—MOMENTS OF INERTIA AND RADII GYRATION.—ALGEBRAICAL EXPRESSIONS FOR SIMPSON'S RULES.—MECHANICAL PRINCIPLES.—CENTRE OF GRAVITY.—LAWS OF MOTION.—DISPLACEMENT, CENTRE OF BUOYANCY.—CENTRE OF GRAVITY OF SHIP'S HULL—STABILITY CURVES AND METACENTRES.—SEA AND SHALLOW-WATER WAVES.—ROLLING OF SHIPS.—PROPULSION AND RESISTANCE OF VESSELS.—SPEED TRIALS.—SAILING, CENTRE OF EFFORT.—DISTANCES DOWN RIVERS, COAST LINES.—STEERING AND RUDDERS OF VESSELS.—LAUNCHING CALCULATIONS AND VELOCITIES.—WEIGHT OF MATERIAL AND GEAR.—GUN PARTICULARS AND WEIGHT.—STANDARD GAUGES.—RIVETED JOINTS AND RIVETING.—STRENGTH AND TESTS OF MATERIALS.—BINDING AND SHEARING STRESSES.—STRENGTH OF SHAFTING, PILLARS, WHEELS, &c.—HYDRAULIC DATA, &c.—CONIC SECTIONS, CATENARIAN CURVES.—MECHANICAL POWERS, WORK.—BOARD OF TRADE REGULATIONS FOR BOILERS AND ENGINES.—BOARD OF TRADE REGULATIONS FOR SHIPS.—LLOYD'S RULES FOR BOILERS.—LLOYD'S WEIGHT OF CHAINS.—LLOYD'S SCANTLINGS FOR SHIPS.—DATA OF ENGINES AND VESSELS.—SHIPS' FITTINGS AND TESTS.—SEASONING PRESERVING TIMBER.—MEASUREMENT OF TIMBER.—ALLOYS, PAINTS, VARNISHES.—DATA FOR STOWAGE.—ADMIRALTY TRANSPORT REGULATIONS.—RULES FOR HORSE-POWER, SCREW PROPELLERS, &c.—PERCENTAGES FOR BUTT STRAPS.—PARTICULARS OF YACHTS.—MASTING AND RIGGING.—DISTANCES OF FOREIGN PORTS.—TONNAGE TABLES.—VOCABULARY OF FRENCH AND ENGLISH TERMS.—ENGLISH WEIGHTS AND MEASURES.—FOREIGN WEIGHTS AND MEASURES.—DECIMAL EQUIVALENTS.—USEFUL NUMBERS.—CIRCULAR MEASURES.—AREAS OF AND CIRCUMFERENCES OF CIRCLES.—AREAS OF SEGMENTS OF CIRCLES.—TABLES OF SQUARES AND CUBES AND ROOTS OF NUMBERS.—TABLES OF LOGARITHMS OF NUMBERS.—TABLES OF HYPERBOLIC LOGARITHMS.—TABLES OF NATURAL SINES, TANGENTS,—TABLES OF LOGARITHMIC SINES, TANGENTS, &c.

"In these days of advanced knowledge a work like this is of the greatest value. It contains a vast amount of information. We unhesitatingly say that it is the most valuable compilation for its specific purpose that has ever been printed. No naval architect, engineer, surveyor, seaman, wood or iron shipbuilder, can afford to be without this work."—*Nautical Magazine.*

NAVAL ARCHITECTURE. An Exposition of the Elementary Principles. By J. PEAKE. Crown 8vo, cloth **3/6**

NAVIGATION AND NAUTICAL ASTRONOMY, In Theory and Practice. By Prof. J. R. YOUNG. Crown 8vo, cloth . . **2/6**
"A very complete, thorough, and useful manual for the young navigator."—*Observatory.*

NAVIGATION, PRACTICAL. Consisting of the Sailor's Sea Book, by J. Greenwood and W. H. Rosser; together with Mathematical and Nautical Tables for the Working of the Problems, by H. Law, C.E., and Prof. J. R. Young **7/0**

PATTERN MAKING. Embracing the Main Types of Engineering Construction, and including Gearing, Engine Work, Sheaves and Pulleys, Pipes and Columns, Screws, Machine Parts, Pumps and Cocks, the Moulding of Patterns in Loam and Greensand, Weight of Castings, &c. By J. G. Horner, A.M.I.M.E. Third Edition, Enlarged. With 486 Illustrations. Crown 8vo, cloth . . . . *Net* **7/6**

"A well-written technical guide, evidently written by a man who understands and has practised what he has written about."—*Builder.*

PATTERN MAKING. A Practical Work on the Art of Making Patterns for Engineering and Foundry Work, including (among other matter) Materials and Tools, Wood Patterns, Metal Patterns, Pattern Shop Mathematics, Cost, Care, &c., of Patterns. By F. W. Barrows. Fully Illustrated by Engravings made from Special Drawings by the Author. Crown 8vo, cloth.
Net **6/0**

PATTERN MAKERS AND PATTERN MAKING—MATERIALS AND TOOLS—EXAMPLES OF PATTERN WORK—METAL PATTERNS—PATTERN SHOP MATHEMATICS—COST, CARE, AND INVENTORY OF PATTERNS—MARKING AND RECORD OF PATTERNS—PATTERN ACCOUNTS.

PIONEER ENGINEERING. A Treatise on the Engineering Operations connected with the Settlement of Waste Lands in New Countries. By E. Dobson, M.Inst.C.E. Second Edition. Crown 8vo, cloth **4/6**

PNEUMATICS, Including Acoustics and the Phenomena of Wind Currents, for the use of Beginners. By Charles Tomlinson, F.R.S. Crown 8vo, cloth **1/6**

PUMPS AND PUMPING. A Handbook for Pump Users. Being Notes on Selection, Construction, and Management. By M. Powis Bale, M.Inst.C.E., M.I.Mech.E. Fifth Edition. Crown 8vo, cloth . **3/6**

"The matter is set forth as concisely as possible. In fact, condensation rather than diffuseness has been the author's aim throughout; yet he does not seem to have omitted anything likely to be of use."—*Journal of Gas Lighting.*

PUNCHES, DIES, AND TOOLS FOR MANUFACTURING IN PRESSES. By Joseph V. Woodworth. Medium 8vo, cloth. 482 pages with 700 Illustrations . . . [*Just Published. Net* **16s.**

SIMPLE BENDING AND FORMING DIES, THEIR CONSTRUCTION, USE AND OPERATION—INTRICATE COMBINATION, BENDING AND FORMING DIES, FOR ACCURATE AND RAPID PRODUCTION—AUTOMATIC FORMING, BENDING AND TWISTING DIES AND PUNCHES, FOR DIFFICULT AND NOVEL SHAPING—CUT, CARRY, AND FOLLOW DIES, TOGETHER WITH TOOL COMBINATIONS FOR PROGRESSIVE SHEET METAL WORKING—NOTCHING PERFORATING AND PIERCING PUNCHES, DIES AND TOOLS—COMPOSITE, SECTIONAL, COMPOUND AND ARMATURE DISK AND SEDGMENT PUNCHES AND DIES—PROCESSES AND TOOLS FOR MAKING RIFLE CARTRIDGES, CARTRIDGE CASES OR QUICK-FIRING GUNS, AND NICKEL BULLET JACKETS—THE MANUFACTURE AND USE OF DIES FOR DRAWING WIRE AND BAR STEEL—PENS. PINS, AND NEEDLES, THEIR EVOLUTION AND MANUFACTURE—PUNCHES, DIES, AND PROCESSES FOR MAKING HYDRAULIC PACKING LEATHERS, TOGETHER WITH TOOLS FOR PAINT AND CHEMICAL TABLETS—DRAWING, RE-DRAWING, REDUCING, FLANGING, FORMING, REVERSING, AND CUPPING PROCESSES, PUNCHES AND DIES FOR CIRCULAR AND RECTANGULAR AND SHEET-METAL ARTICLES—BEADING, WIRING, CURLING AND SEAMING PUNCHES AND DIES FOR CLOSING AND ASSEMBLING OF METAL PARTS—JEWELLERY DIE-MAKING. EYE GLASS LENS AND MEDAL DIES, AND CONSTRUCTION OF SPOON AND FORK-MAKING TOOLS—DESIGN, CONSTRUCTION AND USE OF SUB-PRESSES AND SUB-PRESS DIES FOR WATCH AND CLOCK WORK AND ACCURATE PIERCING AND PUNCHING—DROP FORGING AND DIE SINKING, TOGETHER WITH MAKING OF DROP DIES, STEAM-HAMMER DIES, NUMBER-PLATE TOOLS AND DIES FOR BOLT MACHINES—METHODS, DESIGNS, WAYS, KINKS, FORMULAS AND TOOLS FOR SPECIAL WORK, TOGETHER WITH MISCELLANEOUS INFORMATION OF VALUE TO TOOL AND DIE-MAKERS AND SHEET-METAL GOODS MANUFACTURERS—SPECIAL AND NOVEL PROCESSES' PRESSES AND FEEDS FOR WORKING SHEET METAL IN DIES.

RECLAMATION OF LAND FROM TIDAL WATERS.
A Handbook for Engineers, Landed Proprietors, and others interested in Works of Reclamation. By A. BEAZELEY, M.Inst.C.E. 8vo, cloth. *Net* **10/6**

"The work contains a great deal of practical and useful information which cannot fail to be of service to engineers entrusted with the enclosure of salt marshes, and to land-owners intending to reclaim land from the sea."—*The Engineer.*

REFRIGERATING AND ICE-MAKING MACHINERY.
A Descriptive Treatise for the Use of Persons Employing Refrigerating and Ice-Making Installations, and others. By A. J. WALLIS-TAYLER, A.M.Inst.C.E. Third Edition, Enlarged. Crown 8vo, cloth . . **7/6**

"May be recommended as a useful description of the machinery, the processes, and of the facts, figures, and tabulated physics of refrigerating."—*Engineer.*

REFRIGERATION AND ICE-MAKING POCKET BOOK.
By A. J. WALLIS-TAYLER, A.M.Inst.C.E. Author of "Refrigerating and Ice-making Machinery," &c. Fourth Edition. Crown 8vo, cloth. *Net* **3/6**

REFRIGERATION, COLD STORAGE, & ICE-MAKING:
A Practical Treatise on the Art and Science of Refrigeration. By A. J. WALLIS-TAYLER, A.M.Inst.C.E., Author of "Refrigerating and Ice-Making Machinery." 600 pp., with 360 Illustrations. Medium 8vo, cloth. *Net* **15/0**

"The author has to be congratulated on the completion and production of such an important work and it cannot fail to have a large body of readers, for it leaves out nothing that would in any way be of value to those interested in the subject."—*Steamship.*

RIVER BARS. The Causes of their Formation, and their Treatment by "Induced Tidal Scour"; with a Description of the Successful Reduction by this Method of the Bar at Dublin. By I. J. MANN, Assist. Eng. to the Dublin Port and Docks Board. Royal 8vo, cloth . **7/6**

"We recommend all interested in harbour works—and, indeed, those concerned in the improvements of rivers generally—to read Mr. Mann's interesting work."—*Engineer.*

ROADS AND STREETS. By H. LAW, C.E., and D. K. CLARK, C.E. Revised, with Additional Chapters by A. J. WALLIS-TAYLER, A.M.Inst.C.E. Seventh Edition. Crown 8vo, cloth **6/0**

"A book which every borough surveyor and engineer must possess, and which will be of considerable service to architects, builders, and property owners generally."—*Building News.*

ROOFS OF WOOD AND IRON. Deduced chiefly from the Works of Robison, Tredgold, and Humber. By E. W. TARN, M.A., Architect. Fifth Edition. Crown 8vo, cloth **1/6**

SAFE RAILWAY WORKING. A Treatise on Railway Accidents, their Cause and Prevention; with a Description of Modern Appliances and Systems. By CLEMENT E. STRETTON, C.E. Third Edition, Enlarged. Crown 8vo, cloth **3/6**

SAFE USE OF STEAM. Containing Rules for Unprofessional Steam Users. By an ENGINEER. Eighth Edition. Sewed . **6D.**

"If steam-users would but learn this little book by heart, boiler explosions would become sensations by their rarity."—*English Mechanic.*

CIVIL, MECHANICAL, &c., ENGINEERING. 23

SAILMAKING. By SAMUEL B. SADLER, Practical Sailmaker, late in the employment of Messrs. Ratsey and Lapthorne, of Cowes and Gosport. Second Edition. Revised and Enlarged. Plates. 4to. cloth
[*Just published.* Net. **12/6**

SAILOR'S SEA BOOK. A Rudimentary Treatise on Navigation. By JAMES GREENWOOD, B.A. With numerous Woodcuts and Coloured Plates. New and Enlarged Edition. By W. H. ROSSER. Crown 8vo, cloth **2/6**

SAILS AND SAIL-MAKING. With Draughting, and the Centre of Effort of the Sails. Weights and Sizes of Ropes; Masting, Rigging, and Sails of Steam Vessels, &c. By R. KIPPING, N.A. Crown 8vo, cloth **2/6**

SCREW-THREADS, and Methods of Producing Them. With numerous Tables and complete Directions for using Screw-Cutting Lathes. By PAUL N. HASLUCK, Author of "Lathe-Work," &c. Sixth Edition. Waistcoat-pocket size **1/6**
" Full of useful information, hints and practical criticism. Taps, dies, and screwing tools generally are illustrated and their action described."—*Mechanical World.*

SEA TERMS, PHRASES, AND WORDS (Technical Dictionary (French-English, English-French), used in the English and French Languages. For the Use of Seamen, Engineers, Pilots, Shipbuilders, Shipowners, and Shipbrokers. Compiled by W. PIRRIE, late of the African Steamship Company. Fcap. 8vo, cloth **5/0**

SHIPBUILDING INDUSTRY OF GERMANY. Compiled and Edited by G. LEHMANN-FELSKOWSKI. With Coloured Prints, Art Supplements, and numerous Illustrations throughout the text. Super-royal 4to, cloth . . . . *Net* **10/6**

SHIPS AND BOATS. By W. BLAND. With numerous Illustrations and Models. Tenth Edition. Crown 8vo, cloth . . **1/6**

SHIPS FOR OCEAN AND RIVER SERVICE. Principles of the Construction of. By H. A. SOMMERFELDT. Crown 8vo . **1/6**
ATLAS OF ENGRAVINGS to illustrate the above. Twelve large folding Plates. Royal 4to, cloth **7/6**

SMITHY AND FORGE. Including the Farrier's Art and Coach Smithing. By W. J. E. CRANE. Crown 8vo, cloth . . **2/6**

STATICS, GRAPHIC AND ANALYTIC. In their Practical Application to the Treatment of Stresses in Roofs, Solid Girders, Lattice, Bowstring, and Suspension Bridges, Braced Iron Arches and Piers, and other Frameworks. By R. HUDSON GRAHAM, C.E. Containing Diagrams and Plates to Scale. With numerous Examples, many taken from existing Structures. Specially arranged for Class-work in Colleges and Universities. Second Edition, Revised and Enlarged. 8vo, cloth . . . **16/0**
" Mr. Graham's book will find a place wherever graphic and analytic statics are used or studied."—*Engineer.*

STATIONARY ENGINE DRIVING. A Practical Manual for Engineers in Charge of Stationary Engines. By MICHAEL REYNOLDS, M.S.E. Seventh Edition. Crown 8vo, cloth, **3/6**; cloth boards . **4/6**

"The author is thoroughly acquainted with his subjects, and has produced a manual which is an exceedingly useful one for the class for whom it is specially intended."—*Engineering.*

STATIONARY ENGINES. A Practical Handbook of their Care and Management for Men-in-charge. By C. HURST. Crown 8vo. *Net* **1/0**

STEAM AND THE STEAM ENGINE. Stationary and Portable. Being an Extension of the Treatise on the Steam Engine of Mr. J. SEWELL. By D. K. CLARK, C.E. Fourth Edition. Crown 8vo, cloth **3/6**

"Every essential part of the subject is treated of competently, and in a popular style."—*Iron.*

STEAM AND MACHINERY MANAGEMENT. A Guide to the Arrangement and Economical Management of Machinery, with Hints on Construction and Selection. By M. POWIS BALE, M.Inst.M.E. Crown 8vo, cloth **2/6**

"Gives the results of wide experience."—*Lloyd's Newspaper.*

STEAM ENGINE. A Practical Handbook compiled with especial Reference to Small and Medium-sized Engines. For the Use of Engine Makers, Mechanical Draughtsmen, Engineering Students, and users of Steam Power. By HERMAN HAEDER, C.E. Translated from the German, with additions and alterations by H. H. P. POWLES, A.M.I.C.E., M.I.M.E. Third Edition, Revised. With nearly 1,100 Illustrations. Crown 8vo, cloth . . . . *Net* **7/6**

"This is an excellent book, and should be in the hands of all who are interested in the construction and design of medium-sized stationary engines. . . . A careful study of its contents and the arrangement of the sections leads to the conclusion that there is probably no other book like it in this country. The volume aims at showing the results of practical experience, and it certainly may claim a complete achievement of this idea."—*Nature.*

STEAM ENGINE. A Treatise on the Mathematical Theory of, with Rules and Examples for Practical Men. By T. BAKER, C.E. Crown 8vo, cloth **1/6**

"Teems with scientific information with reference to the steam-engine."—*Design and Work.*

STEAM ENGINE. For the Use of Beginners. By Dr. LARDNER. Crown 8vo, cloth **1/6**

STEAM ENGINE. A Text-Book on the Steam Engine, with a Supplement on GAS ENGINES and PART II. on HEAT ENGINES. By T. M. GOODEVE, M.A., Barrister-at-Law, Professor of Mechanics at the Royal College of Science, London; Author of "The Principles of Mechanics," "The Elements of Mechanism," &c. Fourteenth Edition. Crown 8vo, cloth . **6/0**

"Professor Goodeve has given us a treatise on the steam engine, which will bear comparison with anything written by Huxley or Maxwell, and we can award it no higher praise."—*Engineer.*

"Mr. Goodeve's text-book is a work of which every young engineer should possess himself."—*Mining Journal.*

STEAM ENGINE (PORTABLE.) A Practical Manual on its Construction and Management. For the use of Owners and Users of Steam Engines generally. By WILLIAM DYSON WANSBROUGH. Crown 8vo, cloth **3/6**

"This is a work of value to those who use steam machinery. . . . Should be read by every one who has a steam engine, on a farm or elsewhere."—*Mark Lane Express*.

STEAM ENGINEERING IN THEORY AND PRACTICE. By GARDNER D. HISCOX, M.E. With Chapters on Electrical Engineering. By NEWTON HARRISON, E.E., Author of "Electric Wiring, Diagrams, and Switchboards." 450 Pages Over 400 Detailed Engravings
[*Just published*. *Net* **12/6**

HISTORICAL—STEAM AND ITS PROPERTIES—APPLIANCES FOR THE GENERATION OF STEAM—TYPES OF BOILERS—CHIMNEY AND ITS WORK—HEAT ECONOMY OF THE FEED WATER—STEAM PUMPS AND THEIR WORK—INCRUSTATION AND ITS WORK—STEAM ABOVE ATMOSPHERIC PRESSURE—FLOW OF STEAM FROM NOZZLES—SUPERHEATED STEAM AND ITS WORK—ADIABATIC EXPANSION OF STEAM—INDICATOR AND ITS WORK—STEAM ENGINE PROPORTIONS—SLIDE VALVE ENGINES AND VALVE MOTION—CORLISS ENGINE AND ITS VALVE GEAR—COMPOUND ENGINE AND ITS THEORY—TRIPLE AND MULTIPLE EXPANSION ENGINE—STEAM TURBINE—REFRIGERATION—ELEVATORS AND THEIR MANAGEMENT—COST OF POWER—STEAM ENGINE TROUBLES—ELECTRIC POWER AND ELECTRIC PLANTS.

STONE BLASTING AND QUARRYING. For Building and other Purposes. With Remarks on the Blowing up of Bridges. By Gen. Sir J. BURGOYNE, K.C.B. Crown 8vo, cloth **1/6**

STONE-WORKING MACHINERY. A Manual dealing with the Rapid and Economical Conversion of Stone. With Hints on the Arrangement and Management of Stone Works. By M. POWIS BALE, M.Inst.C.E. Crown 8vo, cloth **9/0**

"Should be in the hands of every mason or student of stonework."—*Colliery Guardian*.

STRAINS, HANDY BOOK FOR THE CALCULATION OF. In Girders and Similar Structures and their Strength. Consisting of Formulæ and Corresponding Diagrams, with numerous details for Practical Application, &c. By WILLIAM HUMBER, A.M.Inst.C.E., &c. Sixth Edition. Crown 8vo, with nearly 100 Woodcuts and 3 Plates, cloth . . . **7/6**

"We heartily commend this really *handy* book to our engineer and architect readers."—*English Mechanic*.

STRAINS ON STRUCTURES OF IRONWORK. With Practical Remarks on Iron Construction. By F. W. SHEILDS, M.Inst.C.E. 8vo, cloth **5/0**

SUBMARINE TELEGRAPHS. Their History, Construction, and Working. Founded in part on WÜNSCHENDORFF'S "Traité de Télégraphie Sous-Marine," and Compiled from Authoritative and Exclusive Sources. By CHARLES BRIGHT, F.R.S.E., A.M.Inst.C.E., M.I.Mech.E., M.I.E.E. Super royal 8vo, nearly 800 pages, fully Illustrated, including a large number of Maps and Folding Plates, strongly bound in cloth
Net **£3 3s.**

"Mr. Bright's interesting written and admirably illustrated book will meet with a welcome reception from cable men."—*Electrician*.

SURVEYING AS PRACTISED BY CIVIL ENGINEERS AND SURVEYORS. Including the Setting-out of Works for Construction and Surveys Abroad, with many Examples taken from Actual Practice. A Handbook for use in the Field and the Office, intended also as a Text-book for Students. By JOHN WHITELAW, Jun., A.M.Inst.C.E., Author of "Points and Crossings." With about 260 Illustrations. Second Edition. Demy 8vo, cloth . . . . *Net* **10/6**

SURVEYING WITH THE CHAIN ONLY—SURVEYING WITH THE AID OF ANGULAR INSTRUMENTS—LEVELLING - ADJUSTMENT OF INSTRUMENTS—RAILWAY (INCLUDING ROAD) SURVEYS AND SETTING OUT—TACHEOMETRY OR STADIA SURVEYING—TUNNEL ALIGNMENT AND SETTING OUT—SURVEYS FOR WATER SUPPLY WORKS—HYDROGRAPHICAL OR MARINE SURVEYING—ASTRONOMICAL OBSERVATIONS USED IN SURVEYING—EXPLANATIONS OF ASTRONOMICAL TERMS—SURVEYS ABROAD IN JUNGLE, DENSE FOREST, AND UNMAPPED OPEN COUNTRY—TRIGONOMETRICAL OR GEODETIC SURVEYS.

"This work is written with admirable lucidity, and will certainly be found of distinct value both to students and to those engaged in actual practice."—*The Builder.*

SURVEYING, LAND AND ENGINEERING. For Students and Practical Use. By T. BAKER, C.E. Twentieth Edition, by F. E. DIXON, A.M.Inst.C.E. With Plates and Diagrams. Crown 8vo, cloth **2/0**

SURVEYING, LAND AND MARINE. In Reference to the Preparation of Plans for Roads and Railways; Canals, Rivers, Towns, Water Supplies; Docks and Harbours. With Description and Use of Surveying Instruments. By W. DAVIS HASKOLL, C.E. Second Edition, Revised with Additions. Large crown 8vo, cloth **9/0**

"This book must prove of great value to the student. We have no hesitation in recommending it, feeling assured that it will more than repay a careful study."—*Mechanical World.*

SURVEYING, PRACTICAL. A Text-book for Students Preparing for Examinations or for Survey Work in the Colonies. By GEORGE W. USILL, A.M.Inst C.E. Eighth Edition, thoroughly Revised and Enlarged by ALEX. BEAZLEY, M.Inst C.E. With 4 Lithographic Plates and 360 Illustrations. Large crown 8vo, **7/6** cloth; or, on thin paper, leather, gilt edges, rounded corners, for pocket use **12/6**

ORDINARY SURVEYING—SURVEYING INSTRUMENTS—TRIGONOMETRY REQUIRED IN SURVEYING—CHAIN SURVEYING—THEODOLITE SURVEYING—TRAVERSING—TOWN-SURVEYING—LEVELLING—CONTOURING—SETTING OUT CURVES—OFFICE WORK—LAND QUANTITIES—COLONIAL LICENSING REGULATIONS—HYPSOMETER TABLES—INTRODUCTION TO TABLES OF NATURAL SINES, ETC.—NATURAL SINES AND CO-SINES—NATURAL TANGENTS AND CO-TANGENTS—NATURAL SECANTS AND CO-SECANTS.

"The first book which should be put in the hands of a pupil of civil engineering."—*Architect.*

SURVEYING TRIGONOMETRICAL. An outline of the Method of Conducting a Trigonometrical Survey. For the Formation of Geographical and Topographical Maps and Plans, Military Reconnaissance. Levelling, &c., with Useful Problems, Formulæ, and Tables. By Lieut.-General FROME, R.E. Fourth Edition, Revised and partly Re-written by Major-General Sir CHARLES WARREN, G.C.M.G., R.E. With 19 Plates and 115 Woodcuts, royal 8vo, cloth **16/0**

SURVEYING WITH THE TACHEOMETER. A Practical Manual for the use of Civil and Military Engineers and Surveyors, including two series of Tables specially computed for the Reduction of Readings in Sexagesimal and in Centesimal Degrees. By NEIL KENNEDY, M.Inst.C.E., With Diagrams and Plates. Second Edition. Demy 8vo, cloth *Net* **10/6**

"The work is very clearly written, and should remove all difficulties in the way of any surveyor desirous of making use of this useful and rapid instrument."—*Nature.*

CIVIL, MECHANICAL, &c., ENGINEERING.

SURVEY PRACTICE. For Reference in Surveying, Levelling, and Setting-out; and in Route Surveys of Travellers by Land and Sea. With Tables, Illustrations, and Records. By L. D'A. JACKSON, A.M.Inst.C.E. Second Edition. 8vo, cloth **12/6**

SURVEYOR'S FIELD BOOK FOR ENGINEERS AND MINING SURVEYORS: Consisting of a Series of Tables, with Rules, Explanations of Systems, and use of Theodolite for Traverse Surveying and plotting the work with minute accuracy by means of Straight Edge and Set Square only; Levelling with the Theodolite, Setting-out Curves with and without the Theodolite, Earthwork Tables, &c. By W. DAVIS HASKOLL, C.E. With numerous Woodcuts. Fifth Edition, Enlarged. Crown 8vo, cloth **12/0**

"The book is very handy; the separate tables of sines and tangents to every minute will make it useful for many other purposes, the genuine traverse tables existing all the same."—*Athenæum.*

TECHNICAL TERMS, English-French, French-English : A Pocket Glossary; with Tables suitable for the Architectural, Engineering, Manufacturing, and Nautical Professions. By JOHN JAMES FLETCHER. Fourth Edition, 200 pp. Waistcoat-pocket size, limp leather . . . **1/6**

"The glossary of terms is very complete, and many of the Tables are new and well arranged We cordially commend the book."—*Mechanical World.*

TOOLS FOR ENGINEERS AND WOODWORKERS. Including Modern Instruments of Measurement. By JOSEPH HORNER, A.M.Inst.M.E., Author of "Pattern Making," &c. Demy 8vo, with 456 Illustrations . . . . *Net* **9/0**

SUMMARY OF CONTENTS :—INTRODUCTION.—GENERAL SURVEY OF TOOLS.—TOOL ANGLES.—SEC. I. CHISEL GROUP.—CHISELS AND APPLIED FORMS FOR WOODWORKERS. —PLANES.—HAND CHISELS AND APPLIED FORMS FOR METAL WORKING.—CHISEL-LIKE TOOLS FOR METAL TURNING, PLANING, &c.—SHEARING ACTION AND SHEARING TOOLS.— SEC. II. EXAMPLES OF SCRAPING TOOLS.—SEC. III. TOOLS—RELATING TO CHISELS AND SCRAPES.—SAWS.—FILES.—MILLING CUTTERS.—BORING TOOLS FOR WOOD AND METAL. —TAPS AND DIES.—SEC. IV. PERCUSSIVE AND MOULDING TOOLS.—PUNCHES, HAMMERS AND CAULKING TOOLS.—MOULDING AND MODELLING TOOLS.—MISCELLANEOUS TOOLS. —SEC. V. HARDENING, TEMPERING, GRINDING AND SHARPENING.—SEC. VI. TOOLS FOR MEASUREMENT AND TEST.—STANDARDS OF MEASUREMENT.—SQUARES, SURFACE PLATES, LEVELS, BEVELS, PROTRACTORS, &c.—SURFACE GAUGES OR SCRIBING BLOCKS. —COMPASSES AND DIVIDERS.—CALIPERS, VERNIER CALIPERS, AND RELATED FORMS.— —MICROMETER CALIPERS.—DEPTH GAUGES AND ROD GAUGES.—SNAP CYLINDRICAL AND LIMIT GAUGES.—SCREW THREAD, WIRE AND REFERENCE GAUGES.—INDICATORS, ETC.

"As an all-round practical work on tools it is more comprehensive than any with which we are acquainted, and we have no doubt it will meet with the large measure of success to which its merits fully entitle it."—*Mechanical World.*

TOOTHED GEARING. A Practical Handbook for Offices, and Workshops. By J. HORNER, A.M.I.M.E. Second Edition, with a New Chapter on Recent Practice. With 184 Illustrations. Crown 8vo, cloth **6/0**

TRAMWAYS: THEIR CONSTRUCTION AND WORKING. Embracing a Comprehensive History of the System; with an exhaustive Analysis of the Various Modes of Traction, including Horse Power, Steam, Cable Traction, Electric Traction, &c.; a Description of the Varieties of Rolling Stock; and ample Details of Cost and Working Expenses. New Edition, Thoroughly Revised, and Including the Progress recently made in Tramway Construction, &c. By D. KINNEAR CLARK, M.Inst.C.E. With 400 Illustrations. 8vo, 780 pp., buckram. **28/0**

TRUSSES OF WOOD AND IRON. Practical Applications of Science in Determining the Stresses, Breaking Weights, Safe Loads, Scantlings, and Details of Construction. With Complete Working Drawings. By W. GRIFFITHS, Surveyor. Oblong 8vo, cloth **4/6**

"This handy little book enters so minutely into every detail connected with the construction of roof trusses that no student need be ignorant of these matters."—*Practical Engineer.*

TUNNELLING. A Practical Treatise, By CHARLES PRELINI, C.E. With additions by CHARLES S. HILL, C.E. With 150 Diagrams and Illustrations. Royal 8vo, cloth . . . . *Net* **16/0**

TUNNELLING, PRACTICAL. Explaining in detail the Setting-out the Works, Shaft-sinking, and Heading-driving, Ranging the Lines and Levelling underground, Sub-Excavating, Timbering and the Construction of the Brickwork of Tunnels. By F. W. SIMMS, M.Inst.C.E. Fourth Edition, Revised and Further Extended, including the most recent (1895) Examples of Sub-aqueous and other Tunnels, by D. KINNEAR CLARK, M.Inst.C.E. With 34 Folding Plates. Imperial 8vo, cloth . . . . . **£2 2s.**

TUNNEL SHAFTS. A Practical and Theoretical Essay on the construction of large Tunnel Shafts. By J. H. WATSON BUCK, M.Inst.C.E., Resident Engineer, L. and N. W. R. With Folding Plates, 8vo, cloth **12/0**

"Will be regarded by civil engineers as of the utmost value and calculated to save much time and obviate many mistakes."—*Colliery Guardian.*

WAGES TABLES. At 54, 52, 50, and 48 Hours per Week. Showing the Amounts of Wages from One quarter of an hour to Sixty-four hours, in each case at Rates of Wages advancing by One Shilling from 4s. to 55s. per week. By THOS. GARBUTT, Accountant. Square crown 8vo, half-bound **6/0**

WATER ENGINEERING. A Practical Treatise on the Measurement, Storage, Conveyance, and Utilization of Water for the Supply of Towns, for Mill Power, and for other Purposes. By CHARLES SLAGG, A.M.Inst.C.E. Second Edition. Crown 8vo. cloth **7/6**

WATER, POWER OF. As Applied to Drive Flour Mills and to give motion to Turbines and other Hydrostatic Engines. By JOSEPH GLYNN, F.R.S., &c. New Edition. Illustrated. Crown 8vo, cloth **2/0**

WATER SUPPLY OF CITIES AND TOWNS. By WILLIAM HUMBER, A.M.Inst. C.E., and M.Inst.M.E., Author of "Cast and Wrought Iron Bridge Construction," &c., &c. Illustrated with 50 Double Plates, 1 Single Plate, Coloured Frontispiece, and upwards of 250 Woodcuts, and containing 400 pp. of Text. Imp. 4to, elegantly and substantially half-bound in morocco . . . . *Net* **£6 6s.**

LIST OF CONTENTS:—I. HISTORICAL SKETCH OF SOME OF THE MEANS THAT HAVE BEEN ADOPTED FOR THE SUPPLY OF WATER TO CITIES AND TOWNS.—II. WATER AND THE FOREIGN MATTER USUALLY ASSOCIATED WITH IT.—III. RAINFALL AND EVAPORATION.—IV. SPRINGS AND THE WATER-BEARING FORMATIONS OF VARIOUS DISTRICTS.—V. MEASUREMENT AND ESTIMATION OF THE FLOW OF WATER.—VI. ON THE SELECTION OF THE SOURCE OF SUPPLY.—VII. WELLS.—VIII. RESERVOIRS.—IX. THE PURIFICATION OF WATER.—X. PUMPS.—XI. PUMPING MACHINERY.—XII CONDUITS.—XIII. DISTRIBUTION OF WATER.—XIV. METERS, SERVICE PIPES, AND HOUSE FITTINGS.—XV. THE LAW AND ECONOMY OF WATER-WORKS.—XVI. CONSTANT AND INTERMITTENT SUPPLY.—XVII. DESCRIPTION OF PLATES.—APPENDICES, GIVING TABLES OF RATES OF SUPPLY, VELOCITIES, &c., &c., TOGETHER WITH SPECIFICATIONS OF SEVERAL WORKS ILLUSTRATED, AMONG WHICH WILL BE FOUND: ABERDEEN, BIDEFORD, CANTERBURY, DUNDEE, HALIFAX, LAMBETH, ROTHERHAM, DUBLIN, AND OTHERS.

"The most systematic and valuable work upon water supply hitherto produced in English, or in any other language. Mr. Humber's work is characterised almost throughout by an exhaustiveness much more distinctive of French and German than of English technical treatises."—*Engineer.*

WATER SUPPLY OF TOWNS AND THE CONSTRUCTION OF WATER-WORKS.
A Practical Treatise for the Use of Engineers and Students of Engineering. By W. K. BURTON, A.M.Inst.C.E., Consulting Engineer to the Tokyo Water-works. Second Edition. Revised and Extended. With numerous Plates and Illustrations. Super-royal 8vo, buckram. **25/0**

I. INTRODUCTORY.—II. DIFFERENT QUALITIES OF WATER.—III. QUANTITY OF WATER TO BE PROVIDED.—IV. ON ASCERTAINING WHETHER A PROPOSED SOURCE OF SUPPLY IS SUFFICIENT.—V. ON ESTIMATING THE STORAGE CAPACITY REQUIRED TO BE PROVIDED.—VI. CLASSIFICATION OF WATER-WORKS.—VII. IMPOUNDING RESERVOIRS.—VIII. EARTHWORK DAMS.—IX. MASONRY DAMS.—X. THE PURIFICATION OF WATER.—XI. SETTLING RESERVOIRS.—XII. SAND FILTRATION.—XIII. PURIFICATION OF WATER BY ACTION OF IRON. SOFTENING OF WATER BY ACTION OF LIME. NATURAL FILTRATION.—XIV. SERVICE OR CLEAN WATER RESERVOIRS—WATER TOWERS—STAND PIPES.—XV. THE CONNECTION OF SETTLING RESERVOIRS, FILTER BEDS AND SERVICE RESERVOIRS.—XVI. PUMPING MACHINERY.—XVII. FLOW OF WATER IN CONDUITS—PIPES AND OPEN CHANNELS.—XVIII. DISTRIBUTION SYSTEMS.—XIX. SPECIAL PROVISIONS FOR THE EXTINCTION OF FIRES.—XX. PIPES FOR WATER-WORKS.—XXI. PREVENTION OF WASTE OF WATER.—XXII. VARIOUS APPLIANCES USED IN CONNECTION WITH WATER-WORKS.

APPENDIX I. By PROF. JOHN MILNE, F.R.S.—CONSIDERATIONS CONCERNING THE PROBABLE EFFECTS OF EARTHQUAKES ON WATER-WORKS, AND THE SPECIAL PRECAUTIONS TO BE TAKEN IN EARTHQUAKE COUNTRIES.

APPENDIX II. By JOHN DE RIJKE, C.E.—ON SAND DUNES AND DUNE SAND AS A SOURCE OF WATER SUPPLY.

"We congratulate the author upon the practical commonsense shown in the preparation of this work. . . . The plates and diagrams have evidently been prepared with great care, and cannot fail to be of great assistance to the student."—*Builder.*

WATER SUPPLY, RURAL.
A Practical Handbook on the Supply of Water and Construction of Water works for small Country Districts. By ALLAN GREENWELL, A.M.Inst.C.E., and W. T. CURRY, A.M.Inst.C.E., F.G.S. With Illustrations. Second Edition, Revised. Crown 8vo, cloth **5/0**

"The volume contains valuable information upon all matters connected with water supply. . . . It is full of details on points which are continually before water-works engineers."—*Nature.*

WELLS AND WELL-SINKING.
By J. G. SWINDELL, A.R.I.B.A., and G. R. BURNELL, C.E. Revised Edition. Crown 8vo, cloth **2/0**

"Solid practical information, written in a concise and lucid style."—*Iron and Coal Trades Review.*

WIRELESS TELEGRAPHY: ITS THEORY AND PRACTICE.
A Handbook for the use of Electrical Engineers, Students, and Operators. By JAMES ERSKINE-MURRAY, D.Sc., Fellow of the Royal Society of Edinburgh, Member of the Institution of Electrical Engineers. Demy 8vo, 338 pages, with over 130 Diagrams and Illustrations.
[Just Published. Net **10/6**

ADAPTATIONS OF THE ELECTRIC CURRENT TO TELEGRAPHY—EARLIER ATTEMPTS AT WIRELESS TELEGRAPHY—APPARATUS USED IN THE PRODUCTION OF HIGH FREQUENCY CURRENTS—DETECTION OF SHORT-LIVED CURRENTS OF HIGH FREQUENCY BY MEANS OF IMPERFECT ELECTRICAL CONTACTS—DETECTION OF OSCILLATORY CURRENTS OF HIGH FREQUENCY BY THEIR EFFECTS ON MAGNETISED IRON—THERMOMETRIC DETECTORS OF OSCILLATORY CURRENTS OF HIGH FREQUENCY—ELECTROLYTIC DETECTORS—THE MARCONI SYSTEM—THE LODGE-MUIRHEAD SYSTEM—THE FESSENDEN SYSTEM—THE HOZIER-BROWN SYSTEM—WIRELESS TELEGRAPHY IN ALASKA—THE DE FOREST SYSTEM—THE POULSEN SYSTEM—THE TELEFUNKEN SYSTEM—DIRECTED SYSTEMS—SOME POINTS IN THE THEORY OF JIGS AND JIGGERS,—ON THEORIES OF TRANSMISSION—WORLD-WAVE TELEGRAPHY—ADJUSTMENTS, ELECTRICAL MEASUREMENTS AND FAULT FINDING—ON THE CALCULATION OF A SYNTONIC WIRELESS TELEGRAPH STATION—TABLES AND NOTES.

WIRELESS TELEGRAPHY; Its Origins, Development, Inventions, and Apparatus. By CHARLES HENRY SEWALL, Author of "Patented Telephony," "The Future of Long-Distance Communication." With 85 Diagrams and Illustrations. Demy 8vo, cloth . . . . *Net* **10/6**

WORKSHOP PRACTICE. As applied to Marine, Land, and Locomotive Engines, Floating Docks, Dredging Machines, Bridges, Ship building, &c. By J. G. WINTON. Fourth Edition, Illustrated. Crown 8vo, cloth **3/6**

WORKS' MANAGER'S HANDBOOK. Comprising Modern Rules, Tables, and Data. For Engineers, Millwrights, and Boiler Makers; Tool Makers, Machinists, and Metal Workers; Iron and Brass Founders, &c. By W. S. HUTTON, Civil and Mechanical Engineer, Author of "The Practical Engineer's Handbook." Seventh Edition, carefully Revised, and Enlarged. Medium 8vo, strongly bound . . . . [*Just published* **15/0**

STATIONARY AND LOCOMOTIVE STEAM-ENGINES, GAS PRODUCERS, GAS-ENGINES, OIL-ENGINES, ETC.—HYDRAULIC MEMORANDA: PIPES, PUMPS, WATER-POWER, ETC.—MILLWORK: SHAFTING, GEARING, PULLEYS, ETC.—STEAM BOILERS, SAFETY VALVES, FACTORY CHIMNEYS, ETC.—HEAT, WARMING, AND VENTILATION—MELTING, CUTTING, AND FINISHING METALS—ALLOYS AND CASTING—WHEEL-CUTTING, SCREW-CUTTING, ETC.—STRENGTH AND WEIGHT OF MATERIALS—WORKSHOP DATA, ETC.

"The volume is an exceedingly useful one, brimful with engineer's notes, memoranda, and rules and well worthy of being on every mechanical engineer's bookshelf."—*Mechanical World.*

PUBLICATIONS OF THE ENGINEERING STANDARDS COMMITTEE.

MESSRS. CROSBY LOCKWOOD and SON, having been appointed OFFICIAL PUBLISHERS to the ENGINEERING STANDARDS COMMITTEE, beg to invite attention to the List given below of the Publications already issued by the Committee, and will be prepared to supply copies thereof and of all Subsequent Publications as issued.

THE ENGINEERING STANDARDS COMMITTEE is the outcome of a Committee appointed by the Institution of Civil Engineers at the instance of Sir John Wolfe Barry, K.C.B., to inquire into the advisability of Standardising Rolled Iron and Steel Sections.

The Committee as now constituted is supported by the Institution of Civil Engineers, the Institution of Mechanical Engineers, the Institution of Naval Architects, the Iron and Steel Institute, and the Institution of Electrical Engineers; and the value and importance of its labours—not only to the Engineering profession, but to the country at large—has been emphatically recognised by His Majesty's Government, who have made a liberal grant from the Public Funds by way of contribution to the financial resources of the Committee, and have placed at its disposal the services (on the several Sub-Committees) of public officials of the highest standing selected from various Government Departments.

The subjects already dealt with, or under consideration by the Committee, include not only Rolled Iron and Steel Sections, but Tests for Iron and Steel Material used in the Construction of Ships and their Machinery, Bridges and General Building Construction, Railway Rolling Stock, Underframes, Component Parts of Locomotives, Railway and Tramway Rails, Electrical Plant, Insulating Materials, Screw Threads and Limit Gauges, Pipe Flanges, Cement, etc.

These particulars will be sufficient to show the importance to the Trade and Industries of the Empire of the work the Committee has undertaken.

Reports already Published :—

1. **BRITISH STANDARD SECTIONS** (9 lists).—ANGLES, EQUAL AND UNEQUAL.—BULB ANGLES, TEES AND PLATES.—Z AND T BARS.—CHANNELS.—BEAMS. *Net* 1/0
2. **TRAMWAY RAILS AND FISH-PLATES.** *Net* 21/0
3. **REPORT ON THE INFLUENCE OF GAUGE LENGTH.**
   By Professor W. C. UNWIN, F.R.S. *Net* 2/6
4. **PROPERTIES OF STANDARD BEAMS.**
   (*Included in No. 6.*) *Net* 1/0
5. **STANDARD LOCOMOTIVES FOR INDIAN RAILWAYS.** *Net* 10/6
6. **PROPERTIES OF BRITISH STANDARD SECTIONS.**
   Diagrams, Definitions, Tables, and Formulæ. *Net* 5/0
7. **TABLES OF COPPER CONDUCTORS AND THICKNESSES OF DI-ELECTRIC.** *Net* 2/6
8. **TUBULAR TRAMWAY POLES.** *Net* 5/0

[P.T.O.

PUBLICATIONS OF THE ENGINEERING STANDARDS COMMITTEE—(continued.)

9. BULL-HEADED RAILWAY RAILS. *Net* **10/6**
10. TABLES OF PIPE FLANGES. *Net* **2/6**
11. FLAT-BOTTOMED RAILWAY RAILS. *Net* **10/6**
12. SPECIFICATION FOR PORTLAND CEMENT. *Net* **2/6**
13. STRUCTURAL STEEL FOR SHIPBUILDING. *Net* **2/6**
14. STRUCTURAL STEEL FOR MARINE BOILERS. *Net* **2/6**
15. STRUCTURAL STEEL FOR BRIDGES AND GENERAL BUILDING CONSTRUCTION. *Net* **2/6**
16. SPECIFICATIONS AND TABLES FOR TELEGRAPH MATERIALS. *Net* **10/6**
17. INTERIM REPORT ON ELECTRICAL MACHINERY. *Net* **2/6**
19. REPORT ON TEMPERATURE EXPERIMENTS ON FIELD COILS OF ELECTRICAL MACHINES. *Net* **5/0**
20. BRITISH STANDARD SCREW THREADS. *Net* **2/6**
21. BRITISH STANDARD PIPE THREADS. *Net* **2/6**
22. THE EFFECT OF TEMPERATURE ON INSULATING MATERIALS. *Net* **5/0**
23. STANDARDS FOR TROLLEY GROOVE AND WIRE. *Net* **1/0**
24. MATERIAL USED IN THE CONSTRUCTION OF RAILWAY ROLLING STOCK. *Net* **10/6**
25. ERRORS IN WORKMANSHIP. Based on Measurements carried out by the National Physical Laboratory. *Net* **10/6**
26. SECOND REPORT ON LOCOMOTIVES FOR INDIAN RAILWAYS . . . . *Net* **10/6**
27. STANDARD SYSTEMS FOR LIMIT GAUGES. (Running Fits) . . . . *Net* **2/6**
28. NUTS, BOLT-HEADS, AND SPANNERS. *Net* **2/6**
29. INGOT STEEL FORGINGS FOR MARINE PURPOSES. *Net* **2/6**
31. STEEL CONDUITS FOR ELECTRICAL WIRING. *Net* **2/6**
32. STEEL BARS (for use in automatic Machines.) *Net* **2/6**
33. CARBON FILAMENT GLOW LAMPS. *Net* **5/0**
34. WHITWORTH, FINE, AND PIPE THREADS. (Mounted on Card and varnished.) *Net* **6D.**

London: Crosby Lockwood & Son,
7 STATIONERS' HALL COURT, E.C.

www.ingramcontent.com/pod-product-compliance
Lightning Source LLC
Chambersburg PA
CBHW031746230426
43669CB00007B/514